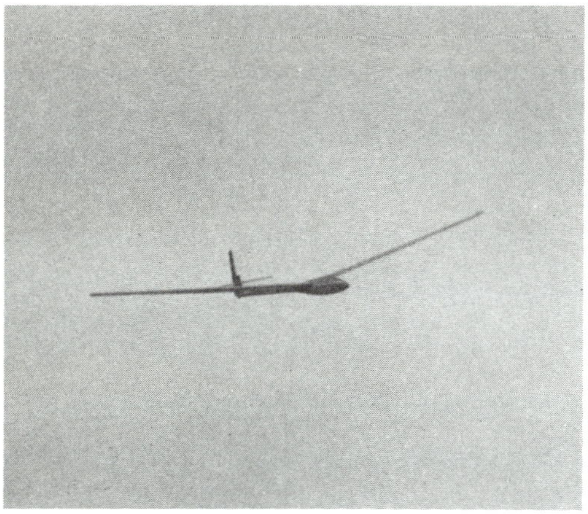

How to build and fly Radio Control Gliders

BY JACK E. SCHRODER

RC PERFORMANCE SERIES NO. 2

Ground School 2

Your First Glider 9

Basic Flying Techniques 17

Aerobatics 24

Clubs, Contests, and Opportunities 28

Useful Addresses 31

Credits and Acknowledgments 32

Editor: Burr Angle Art Director: Lawrence Luser

KALMBACH BOOKS

First printing, 1980. Second printing, 1982. Third printing, 1984. Fourth printing, 1986. Fifth printing, 1987. Sixth printing, 1991.

 Published by Kalmbach Publishing Co., 21027 Crossroads Circle, P.O. Box 1612, Waukesha, WI 53187. Printed in U.S.A. ISBN: 0-89024-549-5.

Ground School

I want you to have fun flying radio control model gliders and I want you to avoid the mistakes of those who learned to fly RC gliders the hard way — by trial and error. I'll help you choose or build your first model, stand beside you as you make your first flights, and introduce you to challenging aerobatic maneuvers. You are probably more interested in flying an RC glider than in reading about aerodynamic or electronic theories, so I'll be as practical as possible. Do read this book from start to finish before you buy a radio system or a glider — you'll get an overview of the entire hobby and save a lot of wasted time and effort.

Lift and drag

Let's begin with brief explanations of lift and drag, two key concepts that recur throughout this book.

A wing produces *lift* because the air pressure on its upper surface is less than that on its lower surface. The pressure of a gas moving over a surface is reduced in proportion to its velocity — the faster the gas moves, the lower the pressure on the surface. The wing of a glider is curved in such a way that air moving past its upper surface travels a greater distance and moves more rapidly than air moving past its lower surface. The difference in pressures is lift. The amount of lift a wing generates depends partly upon the shape (airfoil) of its cross section, partly upon the velocity it moves through the air (airspeed), and partly upon its angle relative to other components of the aircraft (the angle of incidence).

The wing and the rest of the glider are slowed by the friction of the air — *drag.* Think about skipping a flat stone across water: Each time the stone skips it is slowed a little by the drag of the water. When it is slowed enough it stalls and sinks because its speed no longer generates enough lift to keep it skipping. Similarly, every airfoil has a minimum speed below which it ceases lifting and stalls. A large part of your job as an RC glider pilot will be keeping the glider above stalling speed.

Yaw, pitch, and roll

The glider must be oriented so that the wing can produce the most lift with the least drag and so that the ship will fly nose-first. That's why a glider is equipped with vertical and horizontal stabilizers; these work like the feathers on an arrow to control yaw and pitch. *Yaw* is the pivoting of a glider about its vertical axis; the vertical stabilizer ensures that the front edge of the wing (the leading edge) is at a right angle to the oncoming air. *Pitch* is the pivoting of the glider about its lateral axis; the almost equal pressures of the flowing air on the upper and lower surfaces of the horizontal stabilizer cause it to resist upward or downward movement, keeping the wing at

the correct position (angle of attack) to produce maximum lift.

In addition to yawing and pitching, the glider can *roll* — pivot about its longitudinal axis. Dihedral in the wing has major responsibility for preventing unwanted roll. Dihedral means that the tip of each half of the wing is higher than the center. This gives the glider a self-leveling capability because the more one wing half inclines from the horizontal, producing less lift, the more the other wing half inclines toward the horizontal, producing greater lift. To provide even greater stability, halfway out from the fuselage or a little more many glider wings bend upward once again. This second upward bend has the official-sounding name of polyhedral.

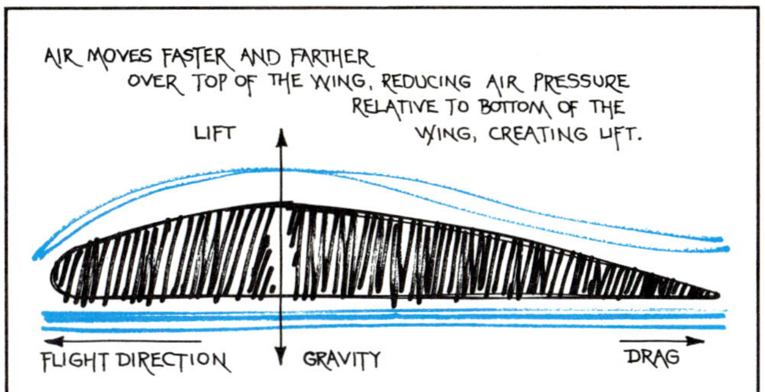

LIFT AND DRAG

Movable control surfaces

We control the flight path of the glider by purposely inducing drag through movements of control surfaces. The *rudder* on the vertical stabilizer turns the nose of the glider left or right. The *elevator* on the horizontal stabilizer raises or lowers the nose, producing climbs and dives. The *ailerons* on the rear edge of the wing (the trailing edge) lift one wing half higher than the other, assisting the rudder in banking and turning the glider. *Spoilers,* which pop up from the upper surface of each wing half, slow the glider by greatly increasing drag when they are deployed.

Man-carrying gliders (full-size gliders) usually have all of these control surfaces, but it's possible to fly an RC glider with rudder only. If the rudder is undeflected the glider flies straight ahead and gains altitude in areas of rising air. If the rudder is deflected to the right (as viewed from the tail), the glider makes a banking right turn, losing altitude as it turns. If the rudder is turned to the left, the glider makes a banking turn to the left, again losing altitude as it turns. The ability to turn and dive the craft is all that an RC pilot needs for simple maneuvers, so rudder-only control is well-suited to model gliders with a wingspan of less than 60″, where very light radio equipment must be used.

Most RC gliders, though, have rudder and elevator controls, as should your first model. Adding an elevator lets you dive and climb the ship without having to turn, and a glider with rudder and elevator can perform most aerobatic maneuvers.

Ailerons allow the pilot to perform advanced aerobatics. They are a luxury for the beginner and complicate learning to fly, so don't consider them until you've mastered the basics. Spoilers are useful for advanced RC pilots whose gliders may fly in very fast-rising air where the spoilers' extra drag helps the pilot keep the craft from gaining altitude too rapidly, but they are definitely not for the beginner.

Automatic dive recovery

A glider is much easier to fly if it is inherently stable, if it "wants" to fly without continuous control commands from the pilot. To help the glider fly straight and to help it recover from a dive, the tips of most glider wings are twisted so that the trailing edge is slightly higher than the leading edge. Air flowing over the wing pushes down on the rear portion of the wing more than the front, helping the glider regain a nose-up attitude whenever the control

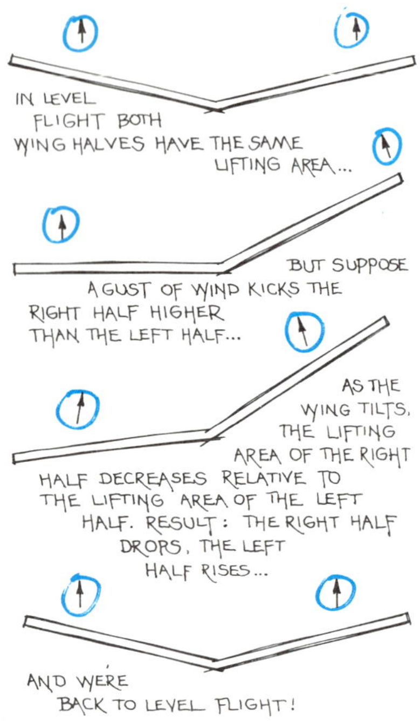

HOW DIHEDRAL WORKS

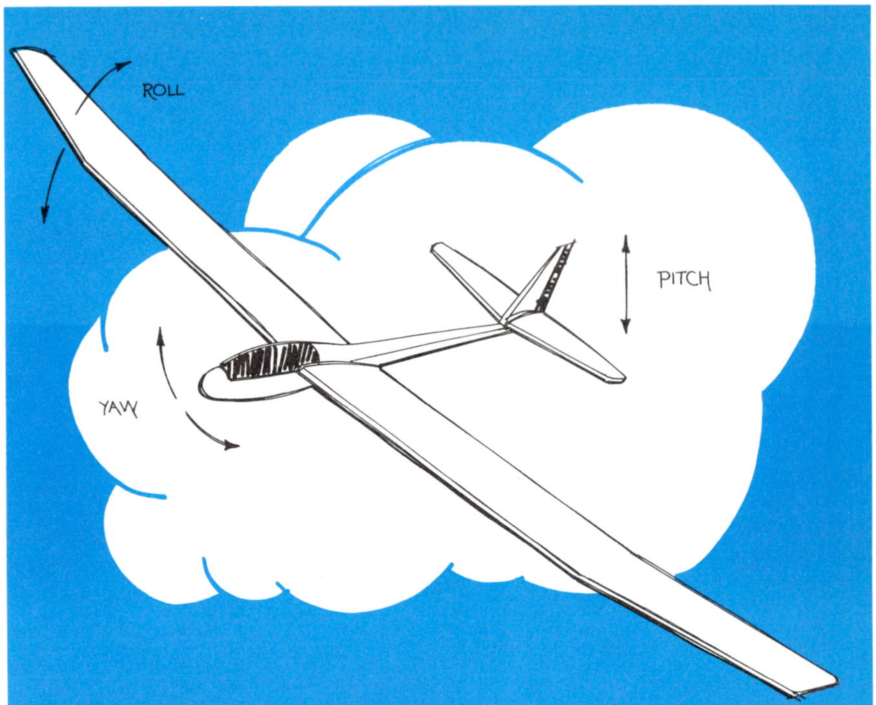

AXES OF ROTATION

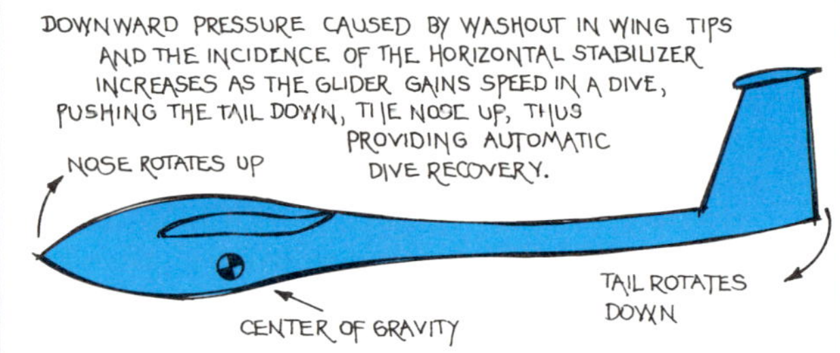

AUTOMATIC DIVE RECOVERY

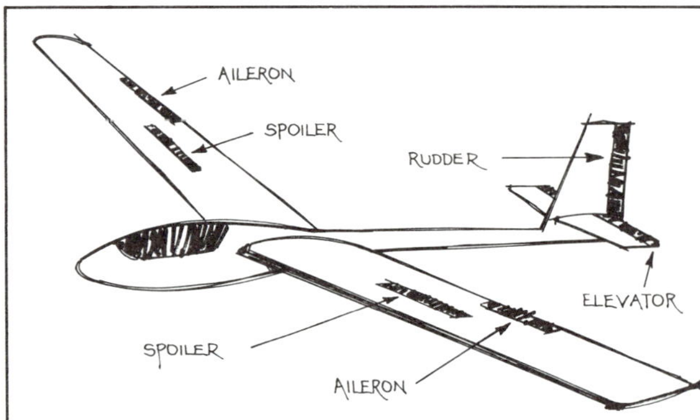

MOVABLE CONTROL SURFACES

surfaces are neutralized (returned to their center position). The upward twist of the trailing edge of the wingtips is *washout*. In addition, the trailing edge of the horizontal stabilizer is usually 2 or 3 degrees higher than the leading edge. The forward movement of the glider causes a downward pressure on the rear portion of the horizontal stabilizer; that pressure (along with washout in the wing tips) keeps the glider's nose up in level flight and helps the glider recover automatically from a dive.

Sailplanes and slope racers

Most RC gliders belong to one of two classes: *sailplanes* and *slope racers*. A sailplane flies slowly in smooth air. A slope racer flies in rough air at high speeds. The different flight environments dictate the design of each type of glider. The sailplane has a long narrow wing designed to produce maximum lift at low speeds (as slow as 10 mph). It is lightly built to ensure the largest possible ratio of forward flight to loss of altitude (glide ratio). The fuselage and other non-lifting parts of the glider are as small as possible in order to keep drag to a minimum. In short, a sailplane is built like a large soaring bird, like an eagle. The slope racer has a shorter, wider wing designed to provide maximum lift at high speeds (100 mph is not uncommon). It is sturdily constructed to withstand the stresses of high-speed aerobatic maneuvers. It is built like a falcon.

Except during launch, a glider is always falling relative to the air in which it is moving. In order to achieve sustained flight a glider must obtain energy to maintain or gain altitude from the air it moves through. A glider remains aloft when its pilot flies the craft in currents of rising air. The two most common forms of rising air used by RC glider pilots are thermal columns and ridge lift.

Thermals

When the surface of the earth is heated unevenly by the sun, the air over the portions that are hotter becomes warmer than the surrounding air and a column of warmer air begins rising through the cooler

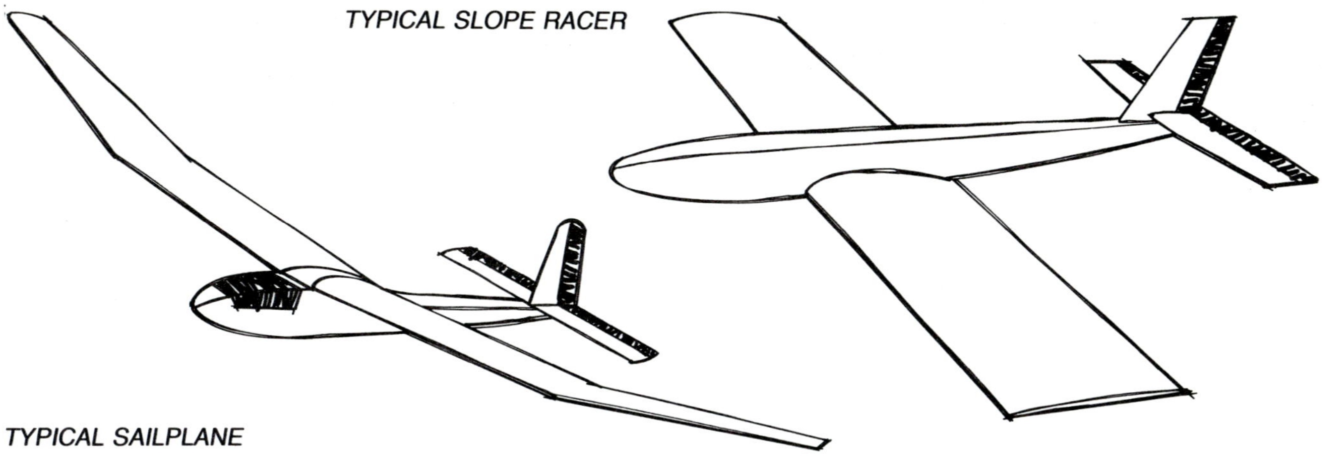

TYPICAL SLOPE RACER

TYPICAL SAILPLANE

STANDING WAVE CAUSED BY A VERTICAL FEATURE

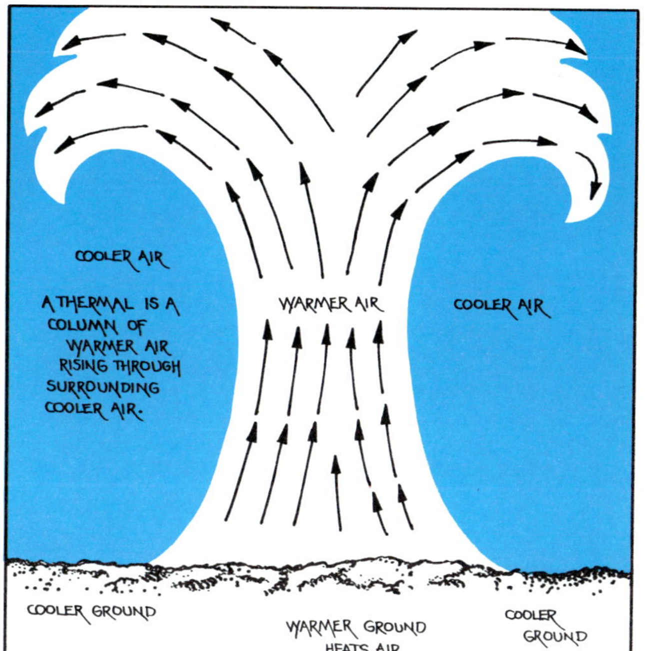

A THERMAL

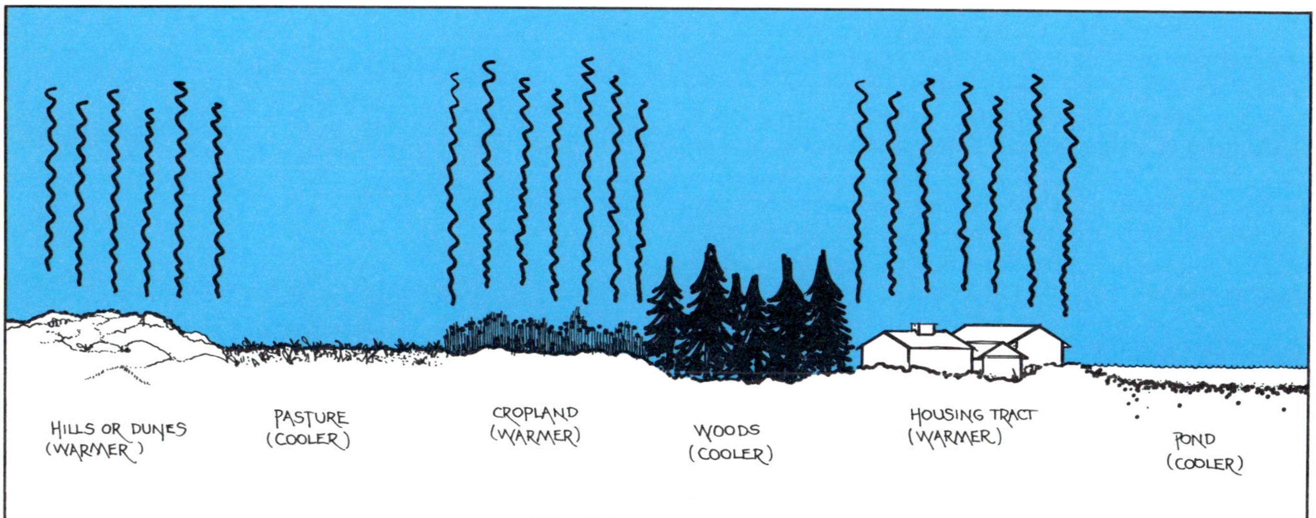

WHERE TO LOOK FOR THERMALS

air. This column of rising air is a *thermal.* A thermal column rises until its temperature is the same as the temperature of the surrounding air. This can be several thousand feet, far higher than you could see even the largest full-size sailplane. Thermals can ascend very rapidly; a South African weather service recorded thermals rising at 8000 feet per minute. That's climbing!

You find thermals by flying over terrain that provides the uneven heating necessary for generating them. Good places to look are dark-colored areas next to light-colored areas such as plowed fields surrounded by crops and pastures. Concrete or blacktop parking lots and roads are great thermal generators. Housing tracts next to large open fields are perfect, especially if the winds blow from the direction of the tract toward the field. Areas of undulating terrain like desert dunes or small foothills are also good.

Thermals are strongest on hot summer days, but you can find workable lift on any day of the year when the sun is shining, and sometimes even when it isn't. The soaring birds fly thermals for hours and hours hunting and just riding herd on their territory. A scientist doing work on the glide of the East African buzzard reported their ability to range hundreds of miles by hitchhiking on thermals. Quite appropriately, he tagged along in a sailplane to observe their behavior.

Some RC glider pilots have thermal-sniffing gadgets such as sensitive anemometers for detecting minute changes in wind velocity and bubble machines for locating areas of rising air. These help, but you can learn to fly RC gliders perfectly well just by watching the birds and doing what they do, always analyzing what's happening and why.

Soaring in thermals requires caution, because RC gliders flown close to disappearing WILL! Always put your name, address, and phone number on your glider in case it does get away and is found by someone in the next county.

Ridge lift

If you live in an area with mountains or hills or ocean cliffs, ridge lift won't be hard to find. Its most common form is the standing wave. The side of a vertical feature facing into the wind forces wind-driven air to pass over its top. As it does so, a wave of steadily rising air, a standing wave, is formed. Another kind of ridge lift is caused by thermal convection. When the sun heats the floor of a canyon or valley, the heated air above the floor expands and forces itself up the sides of the ridge. As it reaches the top, it continues to rise, causing a current of upward-flowing air.

All types of gliders can fly in the rising air furnished by ridge lift, and the sleek slope racers are at their best in this environment. When choosing a spot to fly, be aware that not just any slope will do. The ground in front of the slope should be relatively unobstructed for a mile or two so that the flow of air is not disturbed as it approaches the flying site. Leeward turbulence, the rough air behind the standing wave, can be devastating. Always fly well out in front of the ridge. *Never fly on the side of the hill where leeward turbulence may be encountered.*

Here's an example of the destructive power of

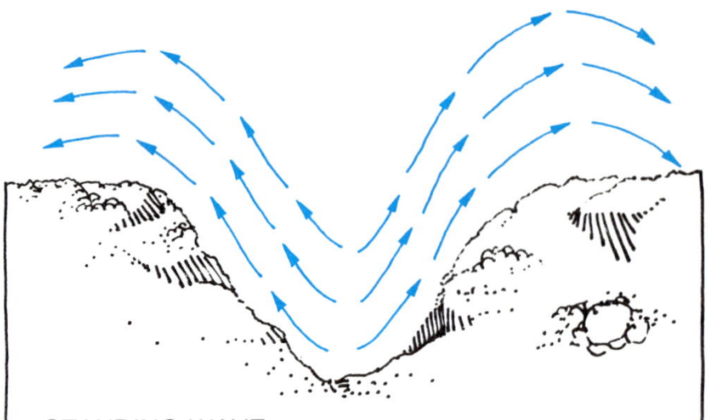

STANDING WAVE CAUSED BY THERMAL CONVECTION

leeward turbulence: While flying a full-size sailplane in the standing wave formed by the Sierra Nevada mountains in California, a pilot carelessly allowed his craft to enter the turbulent area, the rotor, on the downwind side of the range. Huge eddies tore his ship apart, subjecting the pilot to forces 21 times normal gravity. This burst the capillaries in his eyes, causing temporary blindness. After the glider broke up, the pilot pulled his parachute and was sucked back into the rotor where he was banged around for an hour before finally descending near the bottom of the

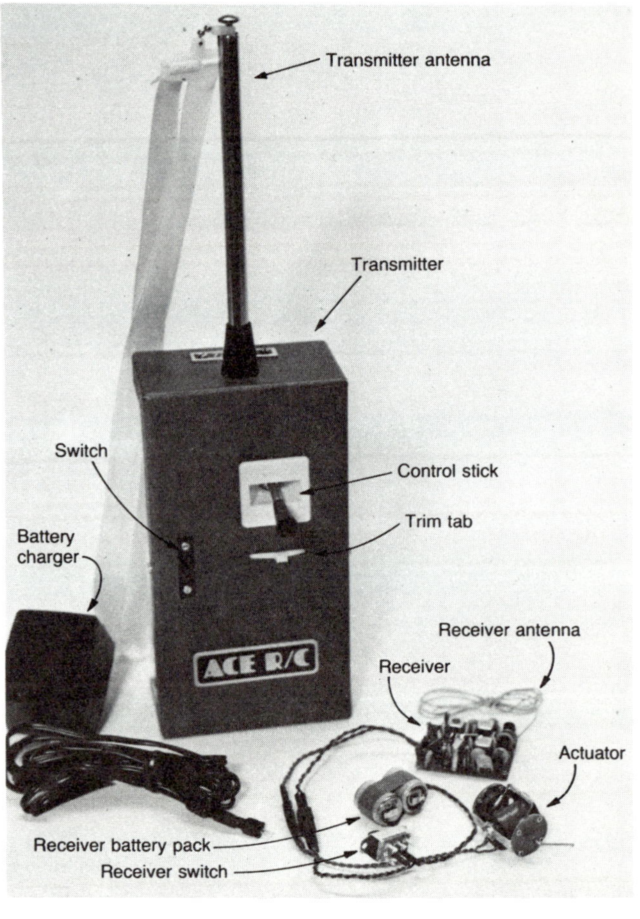

Single-channel radio control system

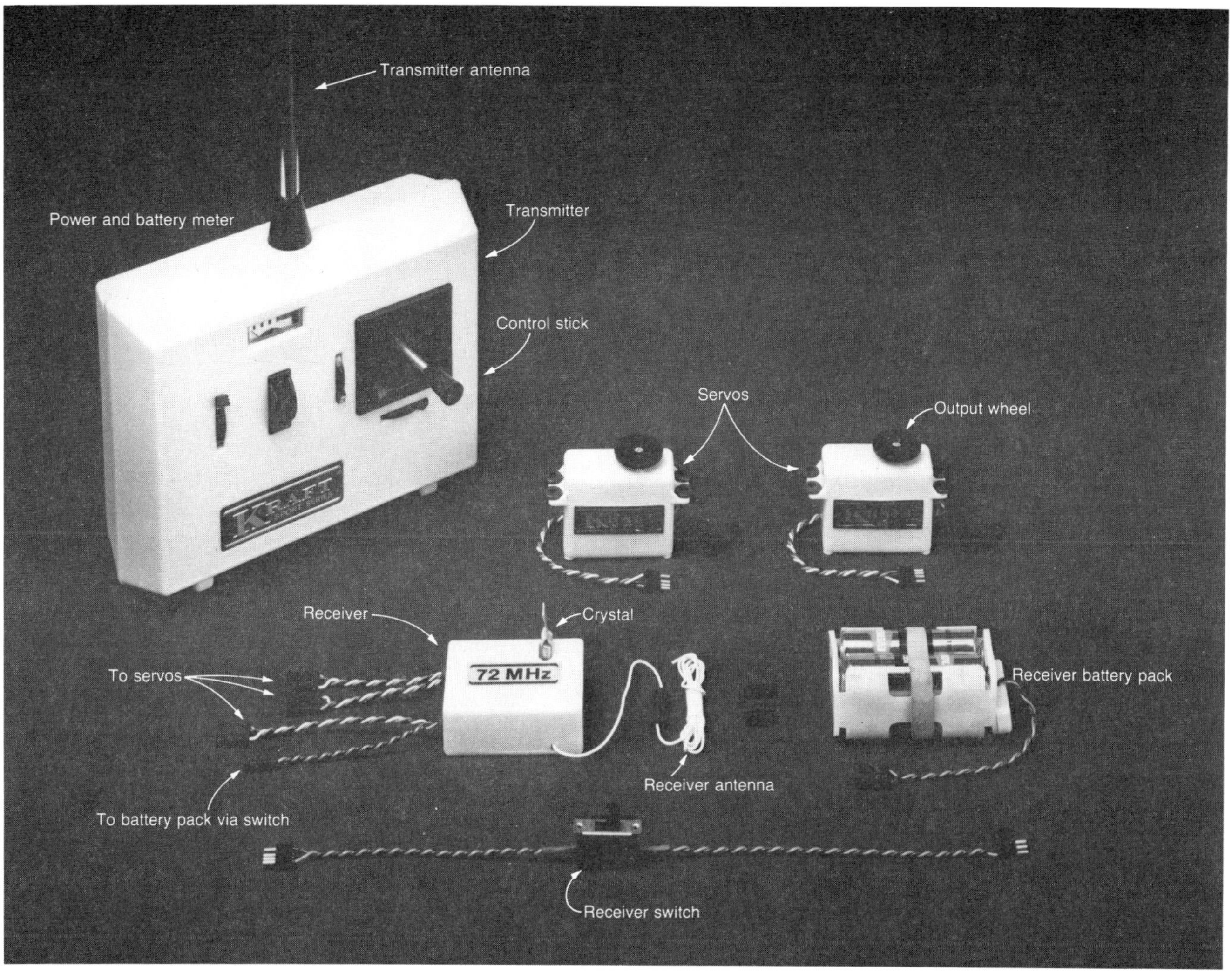

Three-channel radio control system

mountain range. Parts of the sailplane were found as far south as Death Valley, about 100 miles away, and as far east as Nevada, over 250 miles away. What a ride!

Someday you may choose to specialize in flying thermals or ridge lift, but you should start now to evaluate potential flying sites. Look for areas where thermals are likely to form and make note of hills where you can take advantage of ridge lift.

Choosing the RC system

A two or more channel RC system for a glider consists of a transmitter, receiver, servos (short for servomechanisms), batteries, switches, interconnecting cables, and antennas for the transmitter and receiver. The transmitter sends out coded radio signals to the receiver. The receiver decodes the signals and passes the information on to the servos which translate electrical impulses into motion by ordering electric motors to turn in the desired direction. Batteries in the transmitter provide its power source; batteries carried along in the glider power the receiver and servos.

All modern multichannel RC systems are digital and proportional. They transmit information in number-like patterns of on-off pulses — digits — and the motion of the servos in the glider is proportional to the movement of the control stick on the transmitter. Normally one channel is required for each control and one servo is required for each control. Contemporary single-channel RC systems are of a slightly different type — pulse proportional — and use electromagnetic actuators rather than servos, but they operate with the same reliability as digital proportional systems.

How many channels?

RC systems are available with as many as eight channels. As I mentioned, a single-channel system for rudder only is adequate, but a two-channel system for rudder and elevator is better for the beginner because the elevator makes it easier to climb and dive.

If you intend to fly powered RC planes as well as gliders or plan to build gliders with extras like ailerons, spoilers, drag chutes, and retractable landing gear, a four-channel system might be cheaper in the long run. You'll only use two channels in your first glider, but the others will be there when you're ready to move up. As a beginner, stay away from systems with more than four channels unless you want to gizmo yourself into the poorhouse. If you plan on at least moderate involvement with the hobby a three-channel system, which will allow rudder, elevator, and one other control, is a good compromise between starting with too little or too much.

Alkaline cells or nickel-cadmium batteries?

One, two, and three-channel RC systems are usually sold with battery holders but without batteries. You can buy non-rechargeable alkaline AA cells (most often six for the transmitter, four for the receiver and servos) and replace them after several hours of operation. This soon becomes more expensive than buying rechargeable nickel-cadmium batteries and a charger, so I suggest you spend a few extra dollars on your first system and start with nickel-cadmium batteries. Four-channel systems usually come with rechargeable batteries and a charger, so consider this when comparing prices.

Buying your first system

Modeling magazines such as *Model Airplane News* and *RC Modeler* contain advertisements for the many RC systems on the market. Fred Marks, electronics engineer, designer of RC systems, and co-author of BASICS OF RADIO CONTROL MODELING, reports that all major brands perform satisfactorily. Nevertheless, always test equipment before you pay for it; if it doesn't work at the store, there's a good chance it won't work at home either. First, turn on the receiver and transmitter and check that the servos don't chatter and that they don't move unless you move the stick on the transmitter. Then range-check the radio system with the transmitter antenna almost completely collapsed. This will reduce the operating range from several thousand feet to less than 100. You should have complete control up to 75 feet away. If you don't, ask to try another system. The servos must always respond smoothly to any movement of the stick. Any hesitation or slop is unacceptable. Why take chances?

If you are considering used equipment, take or mail it to an FCC-licensed technician for a check-out. You don't want to inherit someone else's trouble and if the system does need service you can get an inexpensive estimate of the repair costs before you decide whether to buy it.

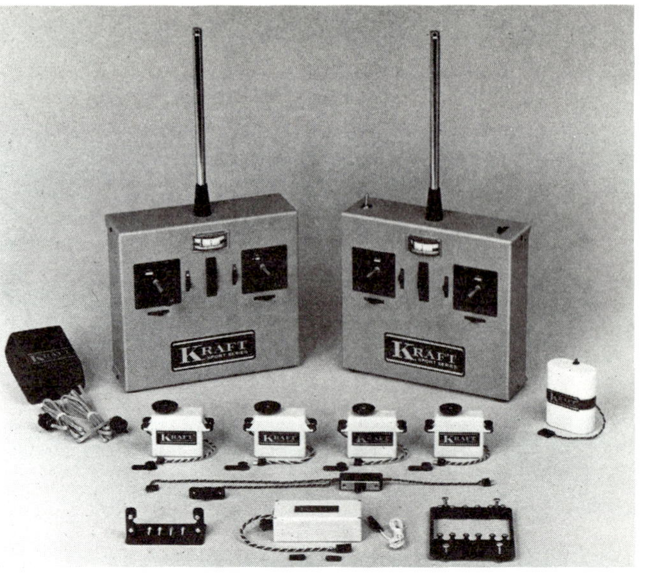

Four- and six-channel radio control systems

Frequencies and Flag Colors

MHz	Color	MHz	Color
26.995	Brown	72.080*	White and brown
27.045	Red	72.160	White and blue
27.095	Orange	72.240*	White and red
27.145	Yellow	72.320	White and violet
27.195	Green	72.400*	White and orange
		72.720*c	
53.1#	Black and brown	72.760*c	
53.2#	Black and red	72.800*c	
53.3#	Black and orange	72.840*c	
53.4#	Black and yellow	72.960	White and yellow
53.5#	Black and green	75.640*	White and green

Amateur radio operators only
* Aircraft only
c Canada only

Frequencies and courtesy

Hobby RC systems used in the United States and Canada operate on frequencies in the 11-meter, 6-meter (amateur radio operators only), and 4-meter bands. The colored flag on the transmitter antenna tells your RC system's frequency to other pilots. Whenever you arrive at a site where other people are operating RC models, always check for your colors on someone else's antenna before you even think about turning on your transmitter. This is because only one glider can use a given frequency at a time — if two transmitters use the same frequency the receivers become confused, leading to crashes and hard feelings.

An FCC Class C citizens band license application, Form 505, should be packed with your RC system. If it isn't, or you've bought used equipment, get the form from the Federal Communications Commission. In Canada, obtain license applications and rules of operation from any office of the Department of Communications. Neither country requires an examination for the license, but both require that you carry the license with you whenever you operate RC models. Perhaps because all early RC modelers were amateur radio operators, the hams' traditions of courteous and lawful use of radio equipment are part and parcel of RC modeling. Abide by FCC or DOC rules at all times.

Care and repair

Modern RC systems use integrated circuits and other technology borrowed from full-size aviation electronics; they are very reliable if treated properly. Proper treatment is mostly common sense — don't drop the components, don't leave them in direct sunlight for hours on end, don't pull hard on cables or antennas, and don't leave the batteries on the charger longer than the manufacturer recommends. If you follow the manufacturer's instructions religiously your RC system should perform satisfactorily for years. Service is available for all currently manufactured systems, so if anything malfunctions or has to be repaired after an especially bad crash, it can be fixed.

Your First Glider

No one reads a few books or watches for a while and becomes an expert RC glider pilot, so your first glider must be durable. Maybe durable is an understatement; let's try practically *indestructible!* There are a number of ready-to-fly, almost-ready-to-fly, and kit-built gliders that fit the bill.

Ready-to-fly and almost-ready-to-fly gliders

If you're in a real hurry to get a ship into the air and don't mind the expense, consider a ready-to-fly model. These come completely assembled by professionals and are almost guaranteed to fly properly because everything is in perfect alignment. You just install your radio gear and go flying. Almost-ready-to-fly models are prefabricated and have compartments in the fuselage for your radio gear; you glue the horizontal and vertical stabilizers squarely on the fuselage and install the radio equipment, jobs that typically take only two to five hours.

Kit-built gliders

Most RC glider pilots, though, build their models from balsa kits. Kit-built gliders take time to assemble, but you will probably enjoy modelbuilding and if you build the model you will know how to repair it after a crash.

There are two glider sizes suitable for learning the basics of RC soaring — approximately 72″ wingspan and approximately 100″ wingspan. Here's a list of 72″ glider kits that meet our criteria of durability, ease of construction, stability, and low cost:

- Airtronics *Square Soar*
- Craft-Air *Drifter II*
- Hobby Shack *Das Slupen Thing*

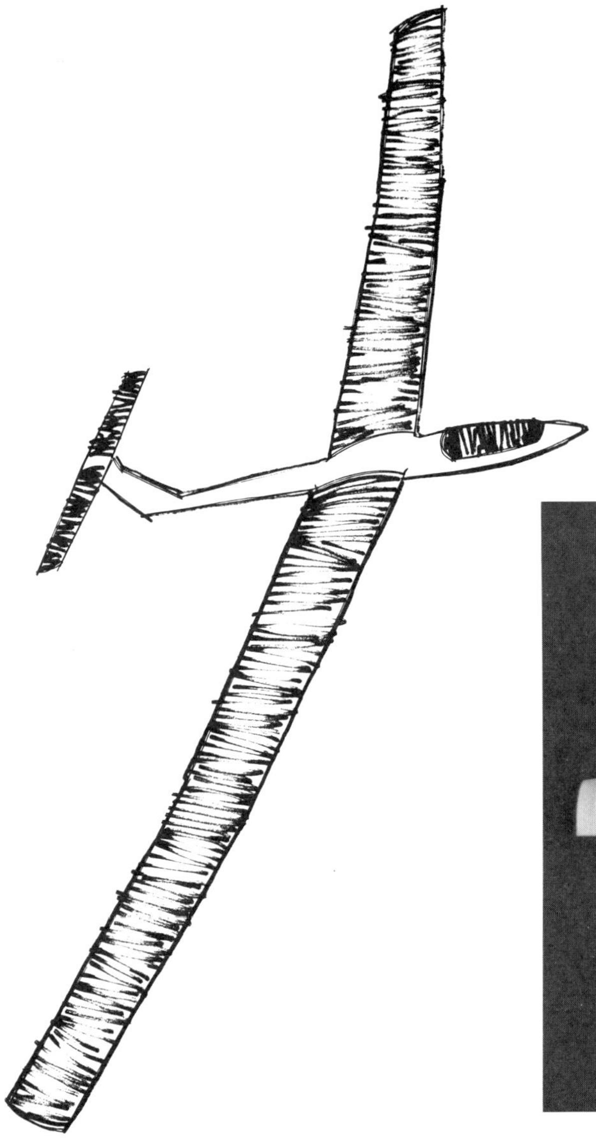

TYPICAL READY-TO-FLY GLIDER

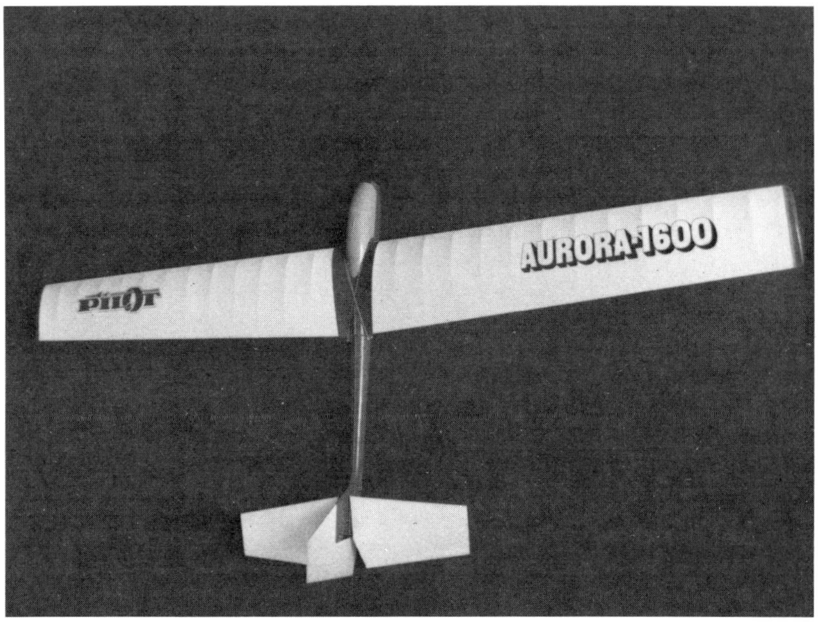

A typical almost-ready-to-fly glider is made almost entirely of plastic. All parts are neatly finished and covered. A few hours' work gluing together the tail and installing the radio gear results in a sturdy glider that flies well.

- Mark's Models *Wanderer 72*
- Pierce Aero *Pierce Arrow*

If carefully constructed, these models will supply all the performance a beginner can handle, but if you want a more impressive ship you can start with one of the 100″ wingspan gliders. These models retain the easy handling characteristics of the smaller ships:

- Airtronics *Aquila*
- Airtronics *Olympic II*
- Astro Flight *Super Malibu*
- Craft-Air *Windrifter*
- Pierce Aero *Pierce Paragon*

These sailplanes are just a few of those available. There are many more, some of which are very beautiful but harder for a novice to build or fly. Because you don't want to get in over your head, stay with the simpler models for your first endeavors. This saves time and money, holds the possibility of construction errors to a minimum, and gets you flying sooner.

Building board

If you decide to build your glider from a balsa kit, first obtain a building board from a lumberyard or building supply store. Any flat board at least 72″ long and 15″ or more wide will do; inexpensive hollow-core or foam-filled door panels usually fit the bill if they are *absolutely flat.* Your building board must be flat so that the wing will come out straight; any twists built into the wing will be impossible to remove, and a warped wing will never fly well. The building board should be 15″ wide to accommodate extra-wide wings should you later want to build a super-size glider.

Tools and materials

When you have selected a good building board, start collecting tools and materials. Most of what you need can be obtained at your local hobby shop; you'll find other items around the house. Obtain a ruler with 1/16″ increments, a box of common pins or T-pins, a small hand drill, an X-acto or other brand of hobby knife, single-edge razor blades, and a razor saw. Waxed paper or Saran Wrap, masking tape, modeler's C-clamps, needle-nose pliers, small screwdrivers, and a small soldering iron and resin core 60/40 electronic solder will also be needed. Always buy the best materials and tools you can afford —

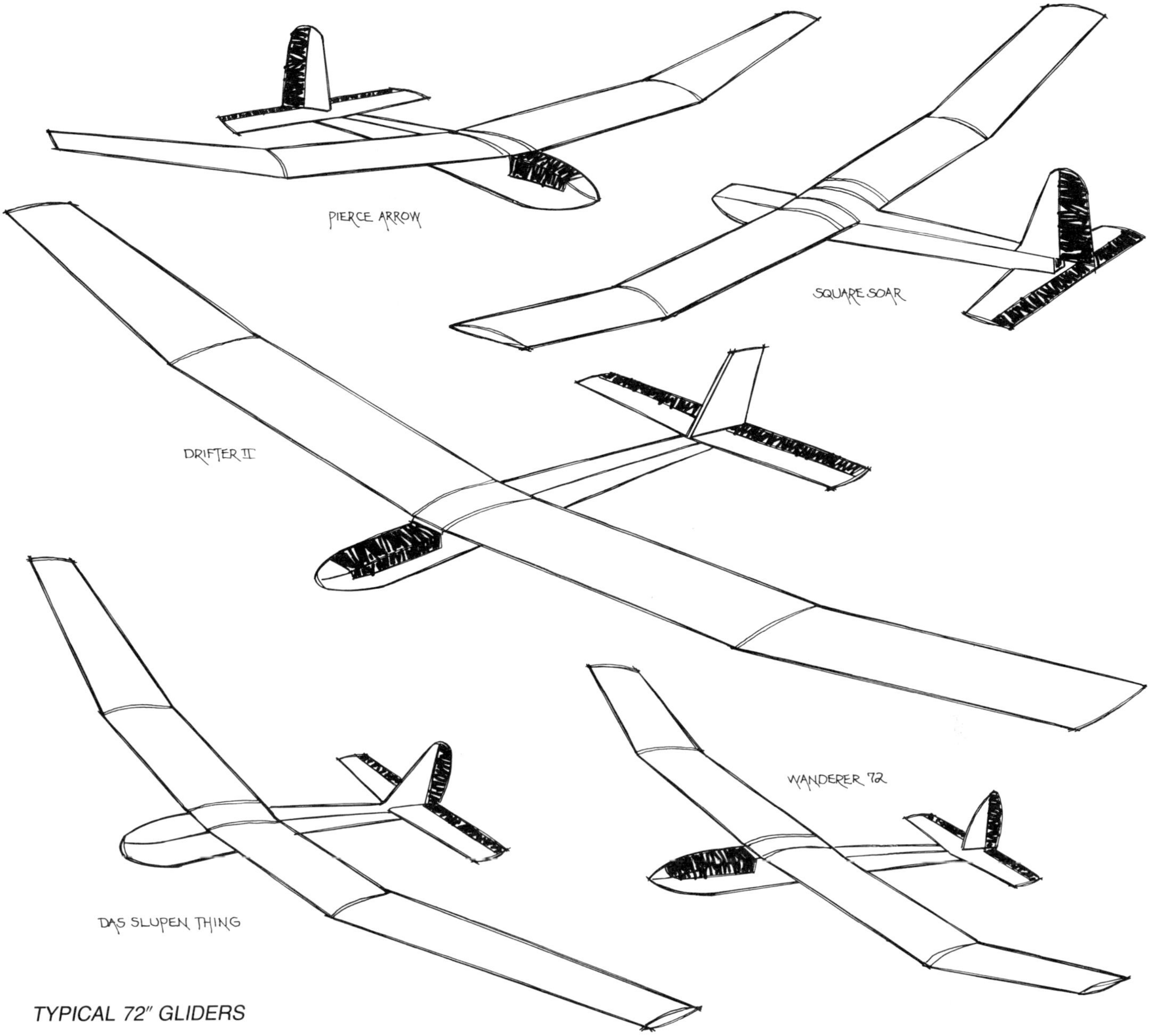

TYPICAL 72″ GLIDERS

they work better and are cheaper in the long run.

Adhesives

Before you buy adhesives study the instruction sheet packaged with your kit and find out what the manufacturer recommends. You will probably do most of your building with wood glue because it is lightweight, easy to work with, and very strong. Aliphatic resin wood glues such as Franklin Titebond or polyvinyl resin white glues such as Elmer's Glue-All are excellent. Use 5-minute epoxy where it is called for on the plans, but apply it sparingly because it is very heavy. Alpha cyanoacrylate cements (ACC's) such as Hot Stuff are great for tacking a piece in place before applying a slow-drying adhesive, but I don't recommend them for the entire construction of the model. ACC's are not as well suited for bonding porous materials as wood glues; if they are used exclusively the structural integrity of the glider may suffer.

Build by subassemblies

You're now ready to begin construction. The glider should be built in three subassemblies — the fuselage, the empennage (the tail section), and the wing. I suggest you build the fuselage first because minor assembly errors there aren't as critical as in the wing. Working on the fuselage first will give you time to get used to the materials and the construction process.

Study the plans and read through the instruction sheet that comes with the kit *twice.* Tape the section of the plan you want to work on to the building board and cover the plan with waxed paper or Saran Wrap so the glue won't stick to it. As you begin assembly, follow the directions step-by-step, sanding or trimming each piece as needed to obtain a perfect fit. If a piece is twisted, steam out the warp or cut a new piece from scrap wood. Check frequently with draftsman's triangles or homemade cardboard triangles to ensure that everything is perfectly square and properly aligned. Use glue sparingly to keep weight down. Your aim is to build the strongest, lightest, best-aligned structure possible.

When one section is assembled, smooth and round it. Rough-sand with a coarse grit sandpaper (No. 180),

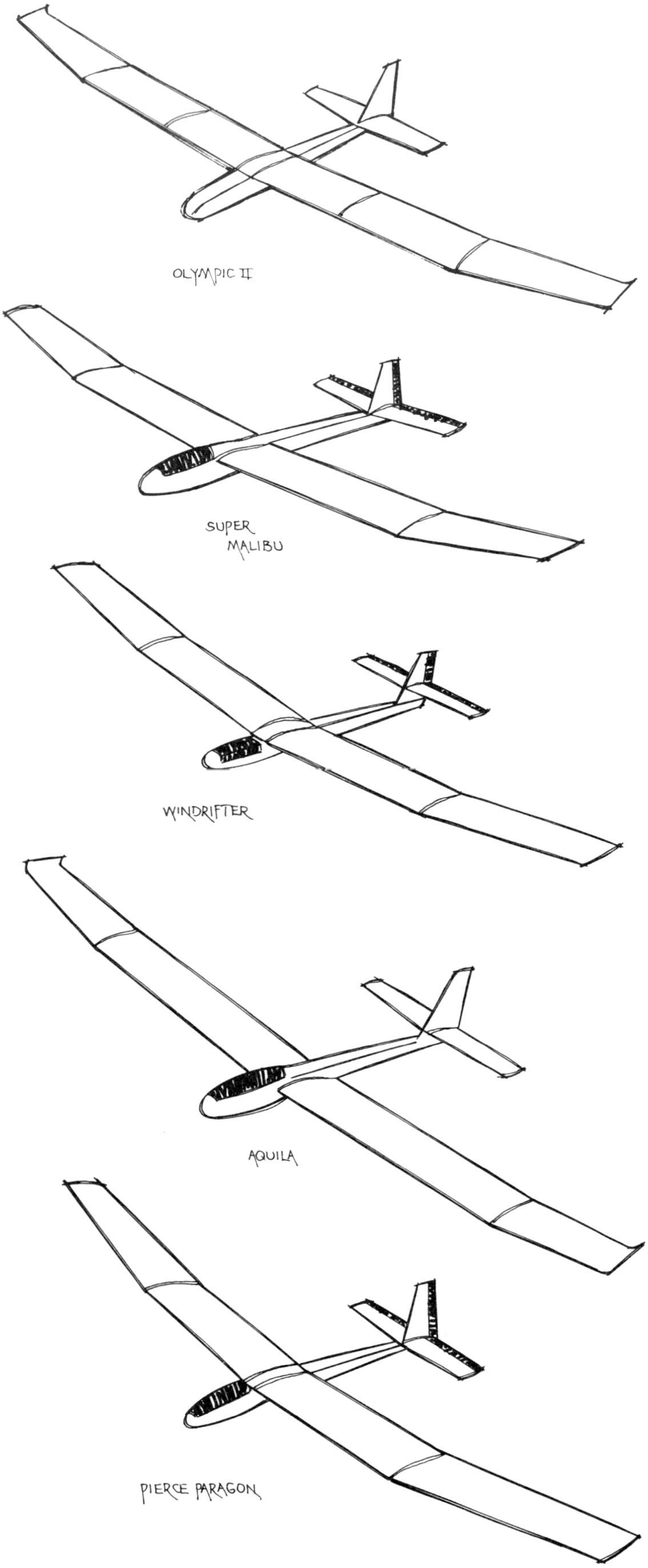

TYPICAL 100″ GLIDERS

and use increasingly finer grits (No. 220, No. 320) down to a finishing paper grit (No. 400). Garnet or wet-or-dry sandpapers work better than flint sandpaper. Always back the sandpaper with a sanding block of scrap wood. Carefully dust and vacuum the surfaces after sanding and put the section aside while you work on another.

Balance the wing

When you build the wing, be especially careful to check for differences in weight of the wood. Take all the parts for one side of the wing and see if they weigh the same as the parts for the other half. If one side is heavier, swap parts from one wing half to the other until they weigh the same. Check the balance again after the wing has been assembled. The wing should remain level when it is turned upside down and balanced on your index finger. If necessary, add small lead weights to the lighter wing half until it balances properly.

Build in correct incidence

After the three main groups have been assembled, sanded, and vacuumed, it's time to check the angles of incidence of the flying surfaces. The wing should have a positive angle of incidence to the imaginary line running fore and aft through the center of the fuselage while the horizontal stabilizer should have a negative angle of incidence. The reason is simple: In flight the wing will be at a 2 or 3 degree angle above horizontal and the stabilizer will be 2 or 3 degrees below horizontal; in a dive pressures on the upper surfaces will pivot the wing up and the tail down, providing (along with washout) an automatic dive recovery system. Adjust the wing and horizontal stabilizer angles of incidence by sanding away excess wood or by gluing in shims of scrap wood until the angles *exactly* match those on the plans.

Install the RC gear

The airborne portion of the radio system consists of the receiver battery pack, receiver on-off switch, receiver, receiver antenna, and the servos. Install these now before covering the model. Mount the battery pack farthest forward, followed by the receiver, then the servos. Wrap the battery pack and the receiver in soft foam plastic (available at your hobby shop) and stuff them into the proper compartments in the fuselage. Use the servo "trays" that came with the RC system (or buy them separately) to mount the servos. Securely glue or screw the servo trays in the places called for on the plans. Mount the switch on the fuselage so that you can turn the receiver on and off without opening any hatches. Lead the receiver antenna through a hole in the side or top of the fuselage; its end will later be attached to the tail. Each piece of airborne RC gear must be securely mounted so that it won't come loose no matter how violently the glider may maneuver or how rapidly it may come to a stop. The heavier parts are placed forward of the lighter ones to help balance the glider.

Install pushrods and control horns

The motion of the servos is translated to control surface movements through a system of pushrods and control horns. One end of the pushrod is attached to

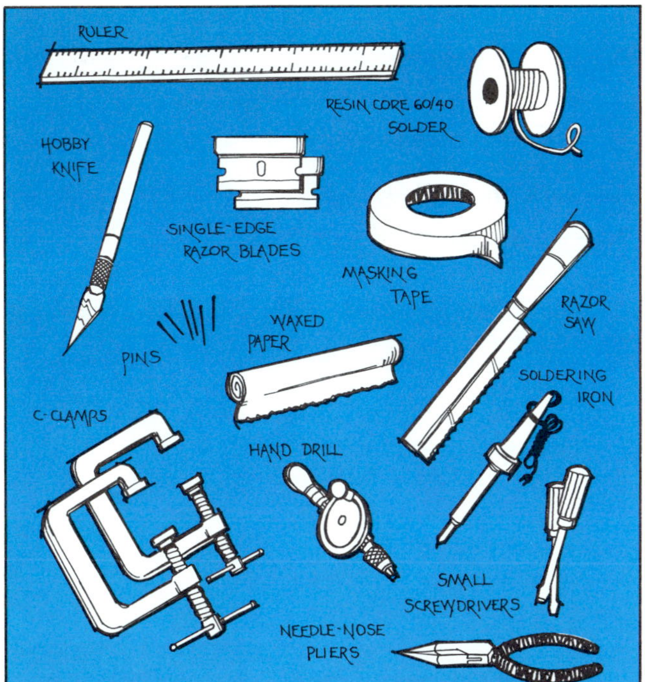

MODELBUILDING TOOLS AND SUPPLIES

the servo output wheel by a clevis and pin device and the other end is connected to a control horn attached to the control surface. If not included with your kit, hardware for these linkages, made by Du-Bro, Goldberg, and others, is available at your hobby shop. The pushrods can be any lightweight rigid material — hard balsa sticks or birch dowels are fine. Control horns are usually made of nylon, polypropylene, or some other tough plastic; attach them exactly as shown on the plans. The hinges connecting the rudder to the vertical stabilizer and the elevator to the horizontal stabilizer are installed by inserting the leaves of the hinges into slots cut in the leading edge of the control surface and the trailing edge of the fixed surface. Work slowly and carefully when installing pushrods, control horns, and hinges because there must be no slop or binding. If the connections are loose, the glider won't respond smoothly to control commands; if the connections are too tight, the glider may not respond to control commands at all, battery life will be shortened, and the servos will wear more rapidly.

Cover the glider

Temporarily remove the battery pack, receiver, and servos before covering the model. As a beginner you are more likely to achieve satisfactory results with plastic covering materials such as Monokote, Solarfilm, and Econokote than with a Silkspan or woven fabric and model airplane dope covering. Experienced modelers achieve beautiful results with fabric and dope (a special lightweight lacquer), but your family may make unpleasant comments about the smell of the dope, and fabric-and-dope covering takes more time than is justified for your first model.

Cover the fuselage first, again because minor mistakes in the fuselage aren't as critical as elsewhere and because the fuselage has lots of sharp curves to deal with, which is good practice before covering the wing. Cut the material to approximate size and tack down the outer edges, then heat the center sections and pull out the wrinkles. If you work with a home iron or hair drier, set the iron on "cotton" or "wool" and the drier at "medium" for Monokote, but use lower temperatures for Solarfilm and Econokote. Experiment with scraps until you find the correct temperature.

After the wing has been covered on both sides, make sure it is still straight. Have a friend carry the wing about 30 feet away and point the trailing edge at you. From this distance it's easy to see if the covering caused any warps. If you find any, twist the wing in the opposite direction of the warp while heating the covering material with the iron or drier. Hold the wing steady while the covering cools. Repeat until all warps are gone.

Twist in washout

Next, place the wing bottom down on your building board. Measure the distance from the trailing edge of each wing tip to the surface of the building board. Raise the trailing edge until you have ⅛″ to ¼″ of washout. Wrinkles will form on the top side of the covering — shrink these out with the iron or drier. Also check the bottom of the wing and remove any wrinkles that formed there. Both wing halves should have exactly the same amount of washout. Don't twist in more washout than is called for on the plans; too much will only cause excess drag.

Balance the glider

Reinstall the radio gear and attach the wing to the fuselage in preparation for balancing the glider. The balance point on most gliders is 30 per cent of the distance back from the leading edge of the wing, close to or at the highest point of the airfoil and on or near the main spar. Your glider's balance point will be clearly marked on the plans. The glider should rest level or slightly nose-down when supported by your fingertips at this balance point. If the tail is lower than the nose, the glider is tail-heavy and the RC gear must be moved farther forward or weight must be added to the nose until the glider balances correctly. This is important: *A tail-heavy glider will not fly properly.* Any ballast called for by the plans must be added as close to the balance point as possible. If the ballast is too far forward the glider will recover from dives too slowly.

Adjust for equal control surface movements

Attach the loose end of the receiver antenna to the vertical or horizontal stabilizer, extend the transmitter antenna, turn on both the receiver and transmitter, stand behind the model, and move the control stick to the right. The rudder should move ¾″ to 1″ to the right. Move the stick to the left and the rudder should move an equal distance to the left. If it doesn't, adjust the pushrod by inserting the pin in the clevis into another hole on the servo output wheel or by screwing the pushrod farther into or out of the clevis. Move the stick forward and the elevator should move down ¾″ to 1″. Move the stick back and the elevator should move up an equal distance. Again, use different output holes and adjust pushrod length until you have attained equal movement.

Step-by-step assembly of a typical kit-built glider

In this photo essay I'm going to take you step-by-step through the assembly of a typical beginner's glider kit. I'll repeat a few things I've said already so that you can't possibly go wrong. Assembly methods vary for different kits, but 90 per cent of the information given here applies to all kit-built balsa gliders.

Open the box and carefully remove the contents. You'll find a set of plans with an instruction sheet or booklet; die-cut or sawed balsa parts for the fuselage, tail, and wing; birch plywood for braces where extra strength is required; and birch dowels for wing holders. The kit may contain hinges for the control surfaces, and hardware for installing the RC gear. It will not include glue or covering material because glue would dry out in storage and because each modelbuilder has his own preferred covering material.

Read the instructions, study the plans, and examine each part until you know where it will go. Spread out the parts as shown in the photo that begins this chapter.

Because we're going to use low-odor and relatively nonflammable materials, your work area can be any well-lighted part of a house or apartment where balsa chips won't be tracked about. Your building board can be any flat surface covered with a material soft enough to allow pins to be driven into it. My building board is a hollow-core door covered with cork tiles.

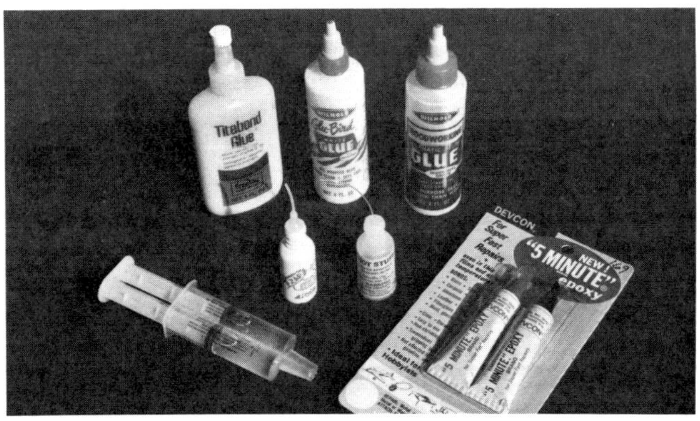

You'll do most gluing with aliphatic resin wood glue such as Titebond and with one of the 5-minute epoxies. The instant-setting alpha cyanoacrylate cements are optional — you may want to use one of them to tack pieces in place before applying wood glue.

Cut out the part of the plan that shows the fuselage, place it on your building board, and cover it with waxed paper.

Place the right and left fuselage sides on the plans and add any required braces. The photos show a Craft-Air *Drifter II,* whose fuselage is reinforced by balsa strips along the fuselage sides. Cut these with a hobby knife fitted with a new blade, coat lightly with wood glue, and pin in place. Push the pins through the balsa into the building board and leave the pieces pinned to the board until the glue dries — at least an hour. Wipe off excess glue with a damp rag before it dries.

Cut the pieces for the fuselage bulkheads (formers), apply glue, pin to the plans, and allow to dry. These

bulkheads must be absolutely square or your model will have a permanently twisted fuselage. Check for squareness with triangles, and keep checking as you assemble all other parts. Glue the bulkheads in place on one side of the fuselage and allow to dry thoroughly. **Now glue the fuselage sides to-**

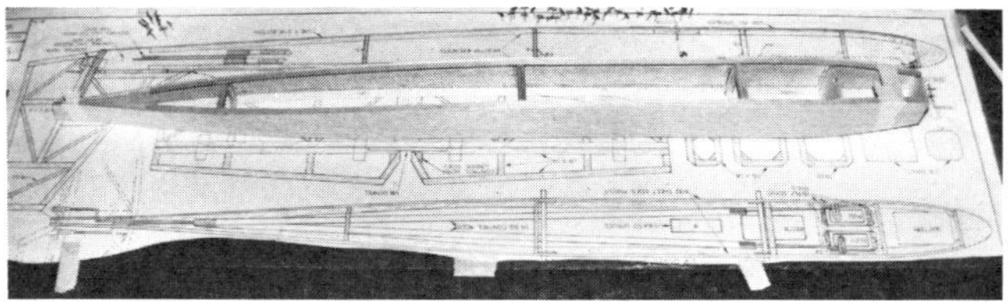

gether using masking tape and pins to hold the pieces until the glue dries. Check with triangles and a long straightedge to ensure perfect symmetry.

Plank the fuselage top and bottom with balsa sheeting applied with the grain running from side to side. Trim these

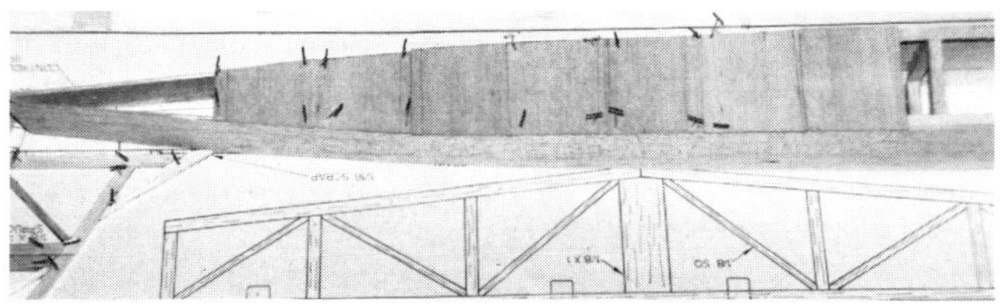

pieces and round the edges with sandpaper backed by a wooden block.

Attach the solid balsa nose block with 5-minute epoxy or wood glue, then carve to shape with a hobby knife. Take small slices and change blades frequently. Balsa is soft, but dulls blades rapidly.

Install the dowels for the rubber bands that hold the wing in place. Prepare hatch covers for the RC compartments and install the towhook. Sand the fuselage smooth. Back the sandpaper with a block to prevent sanding waves into the wood. Dust and vacuum the fuselage and set it aside.

Pin the plans for the vertical and horizontal stabilizers to the building board, cut pieces to size, pin the pieces over the plans, glue, let dry, remove, and sand the edges smooth. Mark the position of the hinges on the vertical and horizontal stabilizers and the rudder and elevator and carefully cut the hinge slots. Use a hobby knife with a new No. 11 blade, and make the slots in the exact center of the parts. **Test-fit the hinges and check for alignment but don't glue them yet.**

Use masking tape, blocks, rulers, and triangles to hold the parts in position and glue the horizontal and vertical stabilizers to the fuselage. The vertical stabilizer must stand straight up and must be in line with the fuselage centerline. The horizontal stabilizer must have the exact angle of incidence called for on the plans. Allow plenty of drying time, then install and glue the hinges to the control surfaces and the vertical and horizontal stabilizers. Use 5-minute epoxy or alpha cyanoacrylate cement.

Examine your work. All joints should be filled with glue, but glue should not have oozed onto other surfaces. If it has, sand it off. Both sides of the fuselage should have the same taper. If they don't, separate the parts with a razor saw and try again. It's better to take the time to correct errors now than to have your first flights ruined by incorrect alignment.

Now mount pushrods and control horns as described in the kit instructions and temporarily install the RC gear. Be prepared to later shift the receiver battery pack forward or back when balancing the glider. Mount the receiver on-off switch on one side of the fuselage so that you won't have to open any hatches to turn the receiver on and off.

Gather up the parts for the wing and swap them around until the parts for each wing half weigh the same. Natural differences in the density of balsa inevitably cause differences in the weight of parts. **Pin the main spar and the leading and trailing edges to the plans and trial fit each rib.** When all

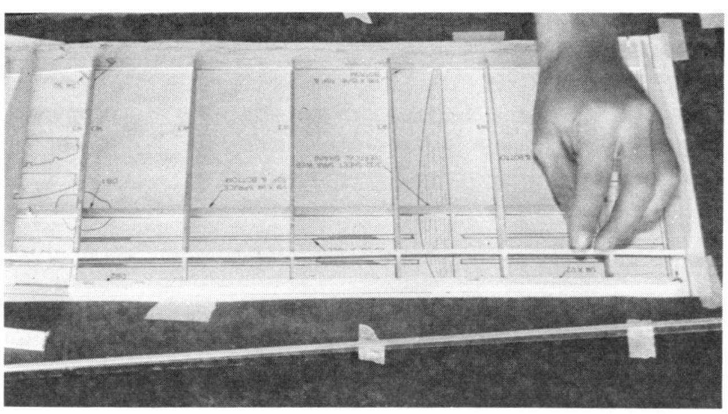

ribs fit perfectly, apply a small amount of wood glue to each joint. Then install the top spars, continually checking for straightness.

The *Drifter II* has a sheet balsa reinforcing section at the center of the wing. Cut the pieces to shape and glue in

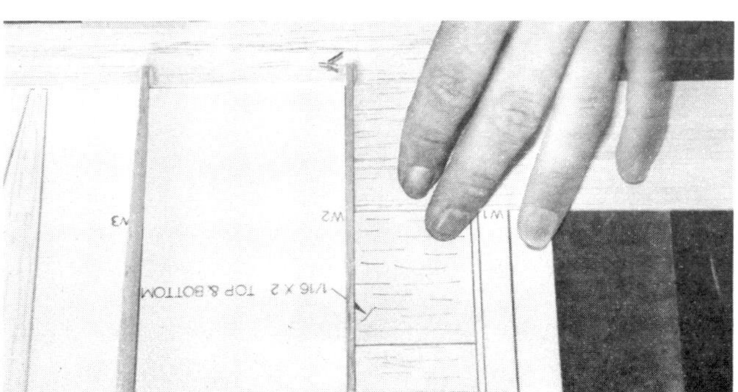

place. Note that the grain runs at a 90 degree angle to the fuselage centerline. If your model has sheet balsa covering the leading portion of the wing, glue this in place now.

Prop up each half of the wing on your building board and use blocks, rolls of tape, or books to raise the ends of the wing to the correct dihedral angle. This angle will be clearly marked in the plans. Carefully measure for perfect alignment, then glue the plywood dihedral braces in place. Repeat this procedure when adding the polyhedral braces in the outer wing sections. Allow these joints to dry thoroughly, then sand. Dust and vacuum the wing.

Attach the wing to the fuselage with rubber bands and check to see if the wing's angle of incidence matches that called for on the plans. If it doesn't, remove excess wood or install balsa shims until the angle is correct.

Fill any dents in the fuselage, tail, or wing with Plastic Wood, spackle, or a homemade mixture of epoxy and microballoons. (Microballoons are tiny phenolic plastic spheres used as a lightweight filler — they are available at your hobby shop.) Apply the filler with a putty or palette knife, let dry, sand, and repeat until all dents and cracks are filled. Use fillers sparingly because they contribute no strength, just weight, to the model.

All that's left to do is covering the model. Buy a roll of Monokote or other brand of plastic covering material at your hobby shop. Once more thoroughly dust and vacuum the model, then proceed:

Step 1. **Remove the backing sheet (if any) and cut the covering material approximately 1″ oversize on all edges.** Use a hobby knife with a new blade or a new single-edge razor blade.

Step 2. Heat your iron to just below the melting point of the covering material. Determine the correct setting by experimenting with scraps. **Gently pull the covering material as smooth as possible and tack down the four corners with the tip of the iron.** Don't worry about the discoloration — it will disappear when the covering cools.

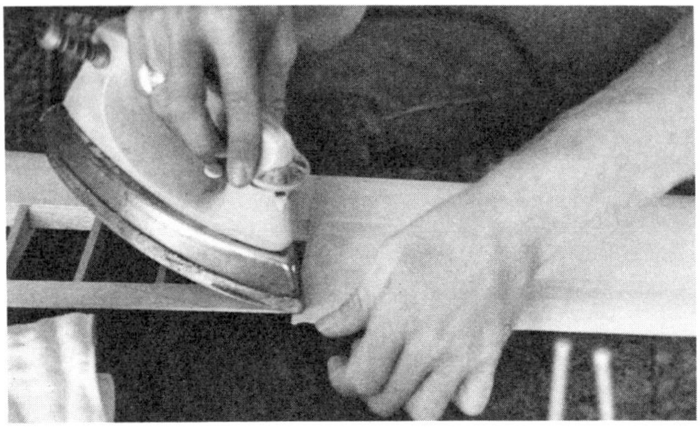

Step 3. Wipe off any adhesive that has oozed onto the covering.

Step 4. **Shrink the material by gently holding the iron against the material.** Do not press hard. Repeat until all of the material is taut.

Step 5. Remove excess covering material with a new single-edge razor blade.

Step 6. **Repeat Steps 2 through 5 on the other surfaces, overlapping the covering ½″.** Trim all edges.

Step 7. **Seal overlapping surfaces and check for warping.**

Now twist in the specified amount of washout, install the RC gear, balance the glider, check for equal control surface movements, and get ready for test flights.

Basic Flying Techniques

Our object now is to make trim adjustments and to get a feel of how the glider behaves in the air. Your first flights will be made from a large, level, unobstructed field (a football field is fine) in absolutely calm air (early morning and early evening are the best times). Take a friend along to help launch the first couple of times so that you can concentrate on controlling the glider.

Trim for straight and level flight

Turn on the radio gear and check for correct operation. If the control surfaces respond correctly, ask your friend to run five yards or so with the glider held aloft to see if it wants to lift. If your friend says the glider doesn't seem to want to fly, don't have him launch it. Adjust the elevator up a little and try again. The correct elevator setting has been attained when the glider, nose pointed at the horizon and wing level, produces a slight upward tug when your launch assistant runs with the glider.

Now have your friend run a few more yards with the glider. As soon as he feels it begin to lift, he should give it a gentle push *straight ahead* and let go. Don't let him throw the glider like a football or baseball. His release should be overhanded, but the nose of the glider must be pointed at the horizon. The glider should fly straight, gradually losing altitude until it lands 30 to 50 feet away. If the glider stalls, add more nose weight until it maintains a level attitude in flight. If the glider noses down steeply, remove nose weight until the glide lengthens. Correct any tendency to turn by adjusting the trim tabs on

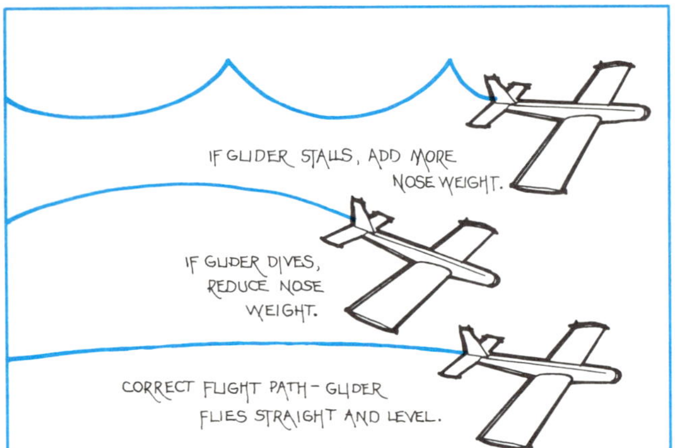

HAND-LAUNCHED TEST FLIGHTS

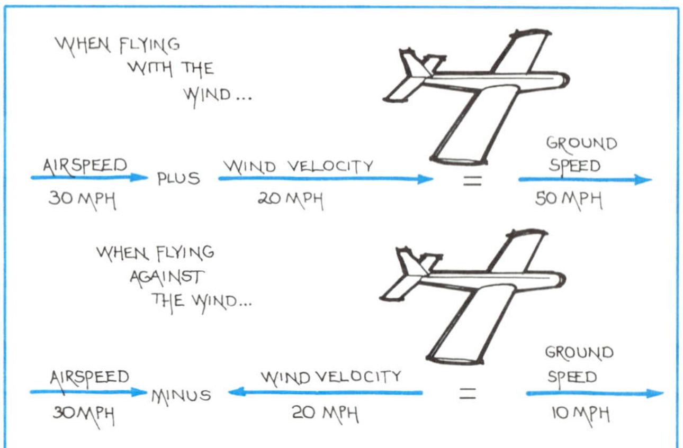

AIRSPEED, WIND VELOCITY, AND GROUND SPEED

the transmitter in the opposite direction. For example, if the glider turns to the left, move the rudder trim tab to the right.

Take your time and make all adjustments in small steps until the plane flies straight and level without using the rudder or elevator controls. Once the glider is trimmed to this "hands off" stable condition, adjust the pushrods as explained in the last chapter so that the control surfaces will remain in their present positions when the trim tabs are returned to neutral.

Low-level practice flights

Now have your friend hand-launch the glider several more times. First, study the way the glider flies on its own, then begin to apply a little elevator or rudder control to learn how the controls affect its flight. Try to land the glider near the same spot each time. Practice making coordinated turns, using rudder to start the turn and up elevator to hold up the nose. You'll discover that only gentle movements of the controls are required — overcontrolling will cause the glider to crash. Return to the field several mornings or afternoons and keep practicing until you can control the glider smoothly.

Downwind turns

Before you venture out of your practice field, let me tell you about the downwind turn, Mother Nature's beast that has claimed more RC gliders than bad radios and midair collisions combined. A glider flying into the wind has a lower ground speed than the same glider flying with the wind, yet the velocity with which the glider must move through the air to avoid stalling remains constant regardless of ground speed. This causes problems for the novice pilot because he equates ground speed (which he can see) with airspeed (which he can't). When he turns the glider downwind, its airspeed drops and it stalls. Stalls on downwind turns can be avoided by maintaining speed during and after the turn, always keeping in mind that ground speed and airspeed must never be confused.

If your glider stalls during a downwind turn, the proper recovery procedure is to apply full opposite rudder while simultaneously moving the stick forward. Follow with neutral rudder and steady back stick until the nose comes up. The object is to first recover from the stall and then to build up enough forward speed to prevent further stalling. Be ready to apply down stick after the glider has regained speed or a second stall may occur.

All downwind turns can be scary because whenever

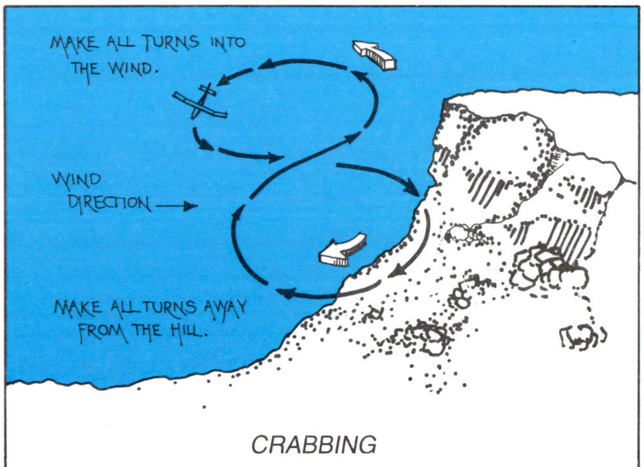

CRABBING

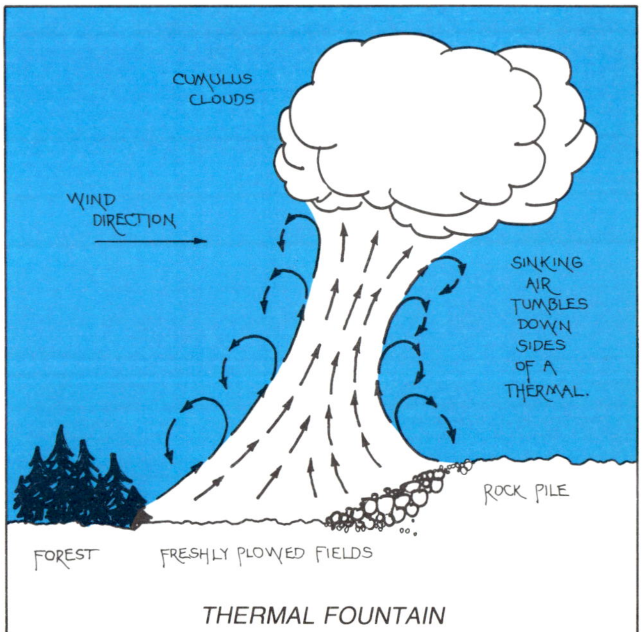

THERMAL FOUNTAIN

the glider flies toward you the controls seem to be reversed. When you push the stick to your right, the glider turns to your left. The glider is, in fact, turning to its right but you may not recognize this until too late. Until you get your bearings you can minimize confusion by always pointing the transmitter antenna in the same direction the glider is moving. Soon you'll imagine yourself in the cockpit of the glider and this won't be necessary.

Confidence versus overconfidence

When you can hand-launch your glider and fly it back to within 10 feet of yourself five times out of six, you've practiced in calm air long enough. Now that you can control the glider in fair conditions and have some confidence you'll really have to resist temptation. Be careful, a novice sailor doesn't enter a transoceanic race just because he can handle a day cruiser. The air has the invisible power to lift a full-size sailplane 2000 feet per minute. If it can do that, you can imagine what might happen to your 2 -pound model.

First altitude flights

A veteran RC glider pilot can give you invaluable advice on where to make your first altitude flights. If

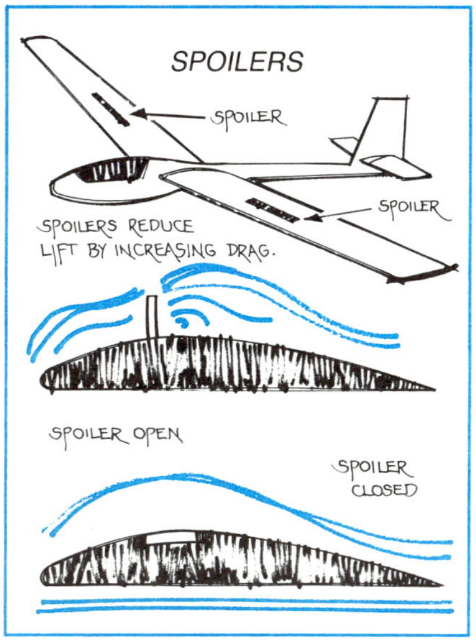

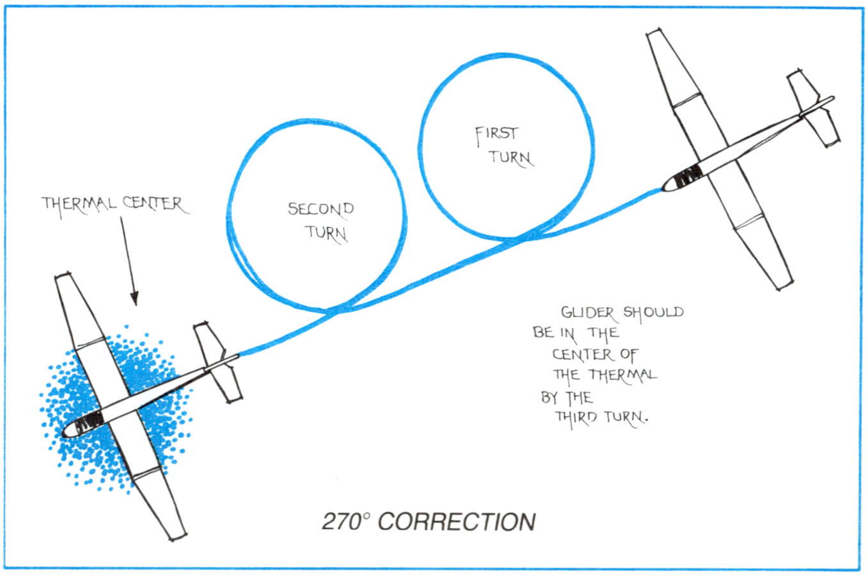

Landings can be tricky. Sometimes even the pilot's body English cannot keep the glider from overturning during a precision landing attempt. At other times, the glider seems to seek out the end of the ribbon which designates the intended landing point.

you can't find an experienced pilot you'll have to search for good flying sites. Folks along the West Coast are lucky because the prevailing westerlies blow in unobstructed off the ocean. This makes for smooth air on the ocean cliffs and provides steady ridge lift in a definite area. If your locale doesn't include cliffs and ocean breezes, mountains, hills, foothills, and even levees often provide just as much lift.

Make your first altitude flights in ridge lift rather than in thermals because you can hand-launch the glider into ridge lift while thermal soaring requires one of the mechanically assisted launching methods explained later in this chapter. Ridge lift is fun no matter where you find it, but as a rule inland ridge lift is more turbulent than coastal ridge lift. Try to find a hill less than 200 feet high that is unobstructed by smaller hills or large buildings for a mile or two in the windward direction.

Begin by hand-launching the glider close to the bottom of the hill. Work up the hill slowly, flying several times from each level. Stay airborne as long as possible and try to land at the same spot each time. Always launch and land into the wind.

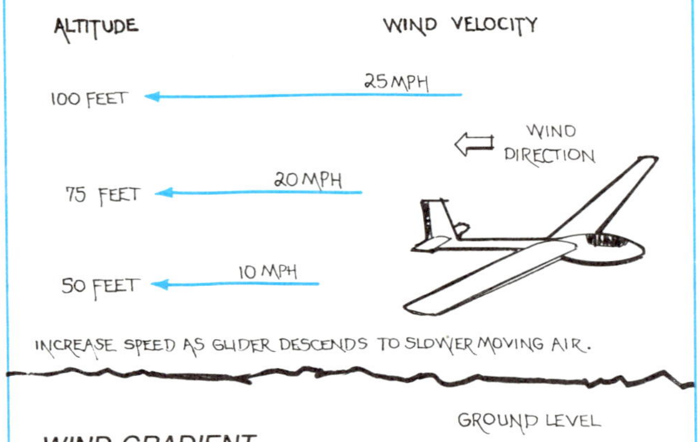

WIND GRADIENT

By the time you work your way to the top of the hill you will notice that the glider gains altitude in certain areas, but loses altitude rapidly in other areas. Fly slowly in the areas of lift and more rapidly where the craft sinks.

After you can maintain control while flying the glider straight into the wind, move on to crabbing. Crabbing is flying parallel to the ridge with the nose of the glider pointed slightly into the wind. Crabbing allows much longer flights.

When working ridge lift the proper speed to fly depends on the glider's altitude. In general, if you are well above the ridge the speed that produces a minimum sink rate is best. If you are closer to the ridge you should fly faster to maintain a reserve of zoom up speed in case of an emergency.

Some other rules to observe while flying ridge lift:

- Make all turns into the wind and away from the slope. This reduces the likelihood of a downwind stall.
- Never fly directly at the slope at minimum airspeed. A bump of turbulent air could stall your glider.
- Give way to the right of an approaching glider. You bank right, so does the other ship.
- Overtake a slower glider by passing between it and the slope so that the slower glider can't suddenly turn in front of you.

Basic thermal flying

If you live in an area where ridge lift does not exist, you'll have to make your first flights using thermal lift. Let's assume that you've launched the glider and it's flying at several hundred feet. Here are a few things about thermals you should be aware of from the start. Thermals work like fountains; hot air bubbles up in the center and spills over the sides as it cools. When you discover a thermal, try to get into the upward-moving air in its center. Once in the center, make a circling turn to the right, being careful to keep the speed high enough to prevent a stall. Stay in the rising air to gain as much altitude as you want.

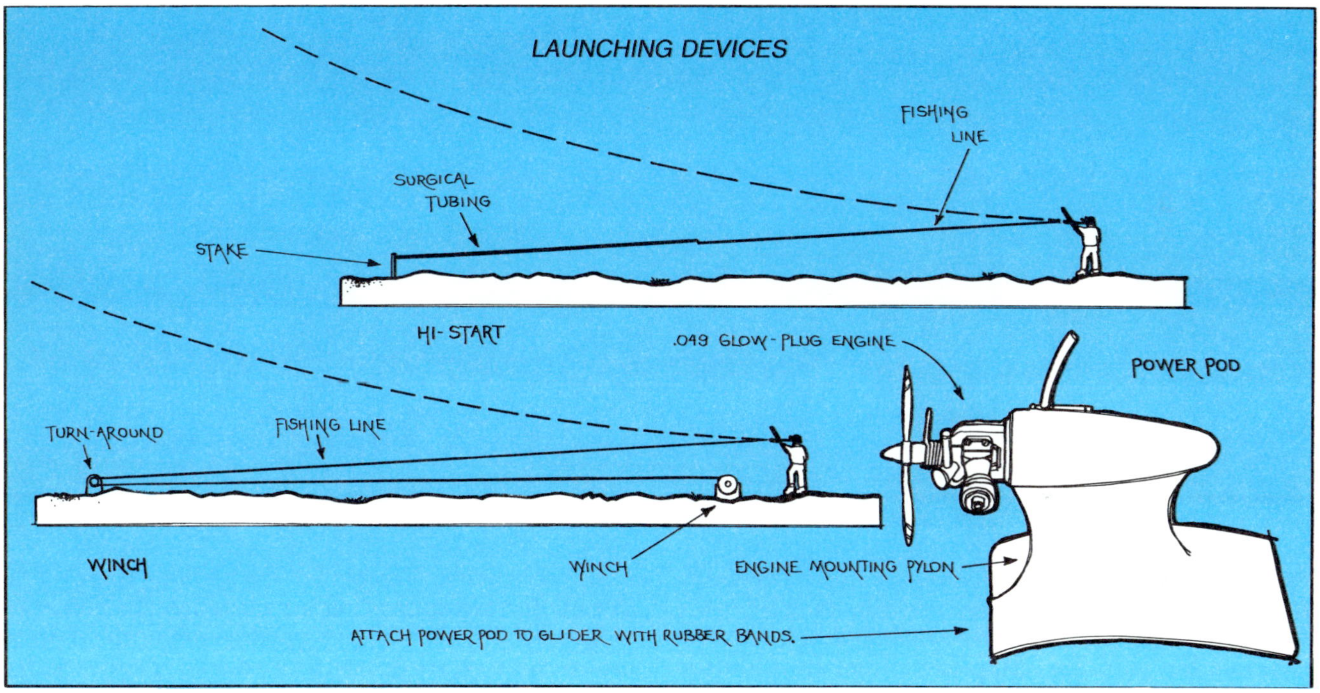

If you get dumped out of the thermal or accidentally turn out of it, make a 270 degree correction to get back in. Start a turn toward the thermal and continue to turn until the glider starts to lift. Then level the wing and fly straight for a few seconds before rolling back into the turn. Don't bank any more steeply than you have to, because you want to expose as much of the wing's lifting surface as possible to the thermal's rising air and banking decreases the lifting area.

As you become more proficient at thermal soaring you'll probably do less flying alone and more in the company of other RC gliders; this brings up the possibility of midair collisions. You will avoid collisions if you follow these rules:

• When your glider enters a thermal already occupied by another sailplane, you must circle in the same direction as the first arrival. If you are gaining altitude faster than the other glider, you must leave it sufficient airspace.

• All diving maneuvers must be done in clear airspace. There is no excuse for diving onto another glider or pulling out of a dive into another craft.

• You must constantly scan the sky for other RC and full-size aircraft. NEVER FLY NEAR FULL-SIZE AIRCRAFT.

Thermals drift downwind as they rise, so get out of your thermal before the glider drifts too far away and fly upwind, searching for another thermal on the way. Particularly on hot summer days, it may be difficult to escape from a fast-rising thermal that threatens to carry your glider out of sight. In such circumstances, straighten out of your last turn, fly upwind, and begin to spiral down at the first hint of zero lift. Be careful not to spiral down so rapidly that the wing is overstressed and torn off. If spiraling doesn't get the ship down, initiate a spin by pulling the stick all the way back and to either side. Once the glider begins to spin, hold the stick in the corner until the desired amount of altitude has been lost, then

Hook up the O-ring, point the nose at the horizon, level the wings, and release, and the hi-start or winch will pull the glider several hundred feet into the air.

move the stick to neutral to recover from the spin. Again, be careful not to overstress the wing.

You won't have them on your first glider, but spoilers are a big help in getting down from strong thermals. At times, though, thermal lift can be so strong that even fully deployed spoilers won't stop your climb. I've flown a full-size sailplane over the California desert during the summer with spoilers deployed only to find myself still climbing at 150 feet per minute! If you get your RC glider into a similar situation, don't panic. Remember that where there's lift, there's downward-moving air close by. It's better to let go of the stick while you plan your escape than to move rashly and compound the problem by overcontrolling.

Landings

A number of things, particularly wind speed and wind direction, must be taken into account when landing. First, be aware that wind velocity varies

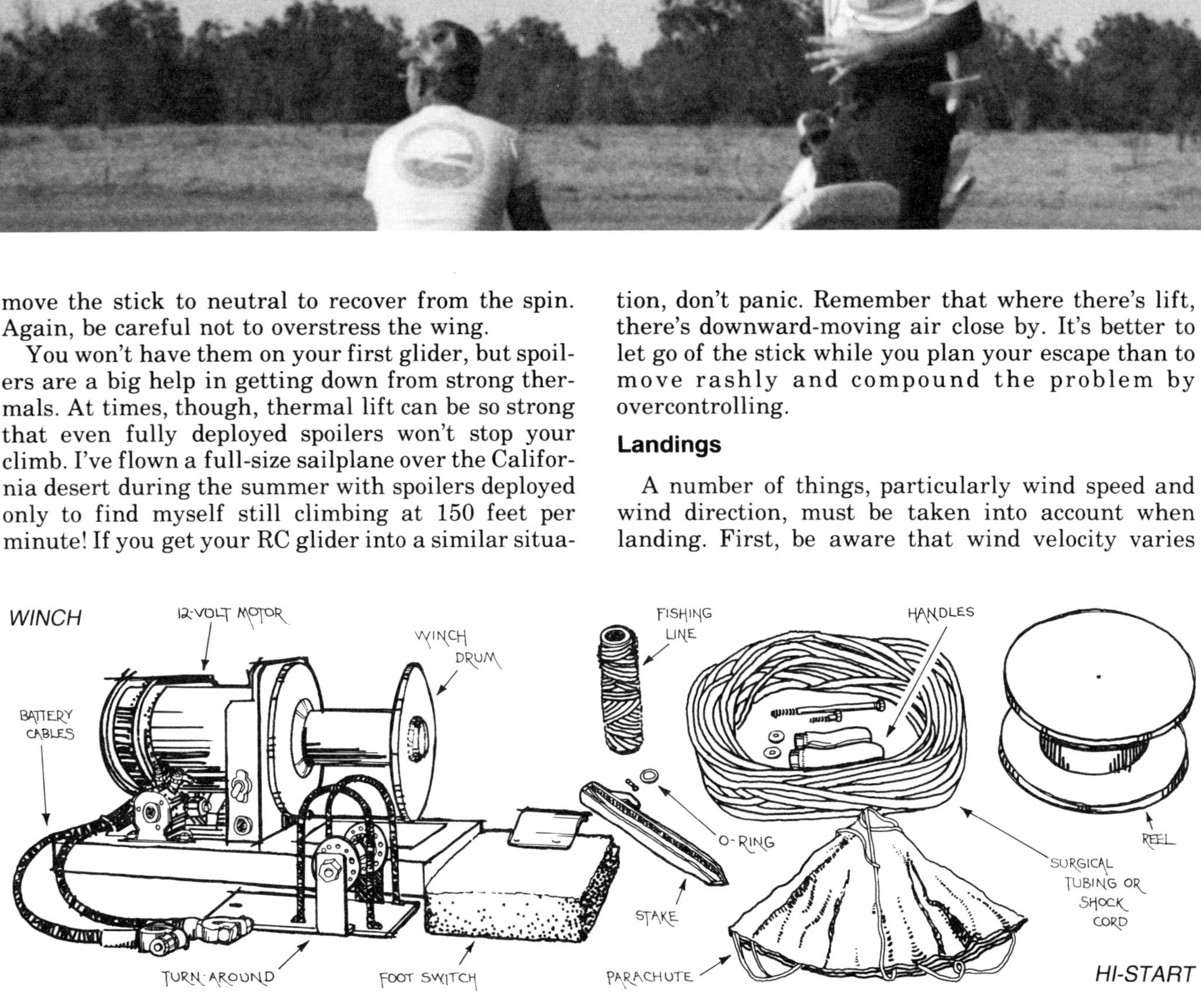

with altitude. Wind speed near the ground is usually less than at various altitudes, and it is never the same. For example, the wind velocity at the surface might be 10 mph, at 50 feet it might be 20 mph, at 75 feet it might be 30 mph. You must increase your glider's airspeed as it loses altitude to compensate for the decreasing wind velocity or the ship will stall. Play safe and keep up speed by setting in a little down elevator trim while still at a safe altitude, say 30 to 50 feet. This way, you'll only have to worry about pulling up the nose on the final landing approach. Also, down elevator trim makes the glider less likely to balloon back into the air when you release the stick after the glider touches down.

Another factor to be taken into account on landing is ground effect. When the glider's altitude is approximately the same as the length of its wing, the wing suddenly seems to produce more lift, lengthening the glide and possibly floating the ship right out of the landing area. If you find yourself trapped by ground effect, make a 90 degree turn and slow the glider until it settles in softly.

The pilot who has to make a downwind landing has made a mistake in judgment somewhere along the line. However, when there's no choice but to make a downwind landing remember the difference between ground speed and airspeed and keep up your airspeed even if the scenery seems to be going by awfully fast. Your choice is between a hotter landing than you would like and a disastrous low-altitude stall. A high-speed landing is less likely to damage the glider than an unrecoverable low-altitude stall, so keep the nose down and try not to hit anything.

Finally, avoid the downwind side of all obstructions. The turbulence on the lee side of a hill or building will vary according to the size and shape of the obstruction and the wind velocity, but turbulence can do your glider no good.

Hi-starts, winches, and power pods

There are three common alternatives to hand-launching your RC glider: hi-start, winch, and power pod. The hi-start is especially useful if you live in western Kansas or some other flat part of the country. It consists of 100 feet of surgical tubing or shock cord connected to 200 to 300 feet of 20- to 50-pound test fishing line with an O-ring at each end. You can make your own or buy one ready-made at a hobby shop. Here's how to use a hi-start: Attach the end of the tubing to a stake in the ground with an O-ring and walk downwind, reeling out the tubing and fishing line as you go. Keep walking until there is approximately 9 pounds of tension on the line and the tubing is stretched to two or three times its original length. Attach the O-ring on the end of the fishing line to a hook installed just forward of the glider's center of gravity. After attaching the ring, ask a friend to hold the glider with both hands, one hand on the nose, one farther back on the fuselage. Stand behind your friend and ask him to point the glider's nose at the horizon with the wing level. Make sure the launch area is clear of birds and aircraft. You are now ready to launch. Turn on and check the radio gear and tell your friend to first release his hand on the nose, then his other hand. The glider will ascend in a graceful arc to several hundred feet. When the tension on the tubing relaxes the ring will fall off the towhook. Start searching for thermals.

It is important not to overcontrol the glider during the tow because a stall may be fatal to the glider. To be safe, use a bit of down trim during launch and trim for level flight after the launch is complete. Always launch into the wind and be certain the glider is sturdy enough to withstand the stresses imposed by the oncoming wind and the tow.

Winches also haul your glider into the air at the end of a long line, but instead of surgical tubing being stretched, 500 to 1000 feet of monofilament fishing line is reeled rapidly onto a winch powered by an electric motor or a small gas engine. Speed is controlled by a foot-operated clutch or on-off switch. Winches are available ready-made or you can build your own with parts from an automobile junkyard. You can launch the glider yourself, but it's best to have a friend operate the winch. Hold and release the glider exactly as when launching with a hi-start. It's better to reel in too slowly than too rapidly because too rapid reeling in may overstress the wing, while reeling in too slowly (as long as the airspeed is above stalling speed) only results in less altitude being gained during launch.

A power pod is a small glow-plug engine (usually an .049) or electric motor temporarily or permanently attached to the glider, most often mounted in a pod over the center of the wing. The power pod lifts the glider to cruising altitude and then shuts off. If you object to the appearance or drag of the pod and propeller, versions are available that retract the pod inside the fuselage or that parachute it to earth. Of course, you don't get the convenience of a power pod for nothing: Glow-plug engines are messy; electric motors, though clean, are heavy.

Aerobatics

Now that you've learned basic flying techniques and are about to try your hand at aerobatics, you must not allow overconfidence to lead you into unsafe flying practices. Every pilot wants to be as good as the best on the hill and he wants to be that good as soon as possible. Plus, most pilots have an almost irresistible urge to show off their aerobatic ability whether they have any or not. Don't you be one of them.

Aerobatics and confidence are great in the proper setting but consider the tale of woe of the hotshot who lost control of his glider and crashed it through a 5x8 foot plate glass window to the tune of $600. After he paid for the window, he had to rebuild his glider to boot. An even worse situation was the case of a guy showing how he could make low, high-speed passes over a group of spectators. A small mistake sent his glider into the crowd, causing a serious injury. By the time the victim was through suing, the cowboy pilot couldn't afford to rebuild his glider or anything else. Never perform aerobatics where spectators could be injured.

And never fly without at least a million dollars' worth of insurance. You may be surprised to learn that courts treat accidents involving model aircraft the same as those involving full-size aircraft. A judge can't take size into account: An aircraft is an aircraft. You can obtain suitable insurance by joining the Academy of Model Aeronautics. AMA membership includes a $1,000,000 insurance policy covering all model flying activities, so I suggest you join now. If you don't join the AMA, you'll have to buy a rider to your homeowner's insurance or find an agent willing to write a special policy.

Low-stress aerobatics

Most RC sailplanes aren't designed for radical aerobatics, and many will break up in midair if high-stress maneuvers are attempted. I'll begin my discussion of aerobatics by explaining low-stress maneuvers that can be performed with a sailplane. Then I'll explain high-stress maneuvers that should be attempted only with a slope racer. You're ready to learn aerobatics when you can consistently keep the glider moving at the same speed and altitude. Practice making faster and faster **figure eights,** pretending there is a pylon at each end, and keep practicing until each turn is clean and steady. This simple maneuver — and all aerobatics — should be performed at least 100 feet from the ground (two mistakes high), but don't go so high that you can't judge the stress on the wing. Flying a glider in strange attitudes can be confusing so think through each maneuver first; planning what to do if a maneuver has to be aborted may save the glider.

Let's try the **inside loop.** Chances are you've already performed this maneuver accidentally, but now I'll tell you how it should be done. If you are flying ridge lift, penetrate out in front of the ridge, maintaining a 100-foot altitude. Point the glider's nose into the wind and steadily push the stick forward until the ship accelerates to slightly more than twice normal flying speed (this should happen within 30 feet of the start of the dive if the dive is steep enough). Unless your glider is unstable in respect to pitch, the nose will come up automatically when you return the stick to neutral; if it doesn't, use a little back stick. As the nose levels, smoothly pull the stick all the way back. When the ship is upside down at the top of the loop first let the stick return to neutral, then pull back smoothly to recover from the dive. At this point the glider will want to loop again so after you've pulled out at the bottom be ready to push the stick forward to keep the nose down.

Each succeeding loop should get smoother until you can loop the glider with perfect symmetry, coming out on a straight heading and maintaining the same speed throughout.

The **chandelle,** a series of high banking figure eight turns, is especially beautiful when performed by sailplanes. To start a chandelle, enter a shallow dive to pick up speed, then allow the nose to come up into a medium climb. As the glider slows, apply rudder in the direction of the desired turn. As the glider skids and slows the nose will drop slightly and the craft will turn 180 degrees. Keep the glider in a shallow dive to regain speed. Push the rudder in the opposite direction and the glider will swing up, outside wing half high, swinging smoothly around at the top of the turn.

The **wingover** is a steeper and slower figure eight turn in which the glider almost comes to a stop at the top of the turn. Enter a climbing turn and apply full rudder at the top of the turn to swing the outside wing half around. When the nose is pointed toward the middle of the eight, it will already be pitched down and the glider will regain speed. Follow the same procedure for the second half of the wingover except use opposite rudder for the opposite turn.

A fourth low-stress aerobatic is the **hammerhead stall.** This maneuver incorporates a full stall. First, dive to build up speed. When the glider is going almost as fast as it will go, smoothly and deliberately bring up the nose to a vertical climb and keep it vertical. As the glider approaches the top of the climb and is slowing to a stop, feed in more and more rudder until at the very top of the climb you have full rudder and the glider stalls, turning 180 degrees. Recovery is the same as for all stalls: Return the controls to neutral, regain flying speed, and pull out of the dive. Be aware that if you initiate a hammerhead stall while the glider is flying away from you, it will be flying toward you when you pull out of the dive. Because of the space required for this maneuver I don't recommend it for low-altitude ridge soaring.

The **spin** is an important functional part of your sailplane's aerobatic repertoire in that it serves well as a dethermalizer. All you have to do to enter a spin is pull the nose up slowly until the glider starts to stall. Right at the point of stall, kick in full up elevator and full rudder to either side. As long as you hold the stick in the lower corner the ship will spin. To recover, let go of the stick, allow the controls to go to neutral, and the glider should come out by itself. If it doesn't, build up flying speed and push the rudder in the opposite direction of the spin to stop the rotation, then smoothly pull back to raise the nose. If this manual recovery method is necessary, the glider has too much nose weight, too much washout, or too much negative incidence in the horizontal stabilizer, so make adjustments until it will recover automatically.

Most sailplanes are not designed to **roll,** but if you're sure your glider can stand the negative loads imposed by the maneuver, you can modify it so it will roll. Enlarge the rudder by 25 per cent, use longer throwing control arms on the servos to increase rudder deflection, or install ailerons. Start the roll by entering a dive to build up speed. Let the nose come up at the bottom of the dive, keeping speed high. When the nose comes past the horizon and starts to climb, immediately apply and hold in full rudder in the direction you want the glider to roll. Use the ailerons to help kick the ship over. When the glider is upside down you may need a little forward stick to keep the nose from dropping. As the glider continues around, pull the stick back and hold full rudder until the maneuver is completed. Be absolutely sure when you enter the roll that the glider is not in a dive and is starting to nose up or the glider may end up upside down in a high-speed spiral dive.

High-stress aerobatics

When you feel like cutting up and it's too windy for thermalling, break out a slope racer and go a little crazy. A 15-mph wind blowing up a cliff or any steep face where standing waves can form is perfect for these speedsters. Outside loops, snap rolls, and other high-stress maneuvers that sailplanes can't do come easily to slope racers.

Outside loops are begun by pushing the stick forward one-third to two-thirds of its total throw. When the ship is moving fast enough, push the stick all the way forward and hold it there until the racer levels out at the top of the loop.

Snap rolls are done by building up speed, leveling out, and pushing in full right or left stick. The racer should snap around very quickly — so quickly that it may start a second roll, so watch out for that. If your racer doesn't roll properly, lengthen the servo output arms or move the clevis on the control horns closer to the base of the horn. This will provide greater control surface movement.

Begin a **Cuban Eight** as if you were performing a loop, but when the glider reaches the top of the loop

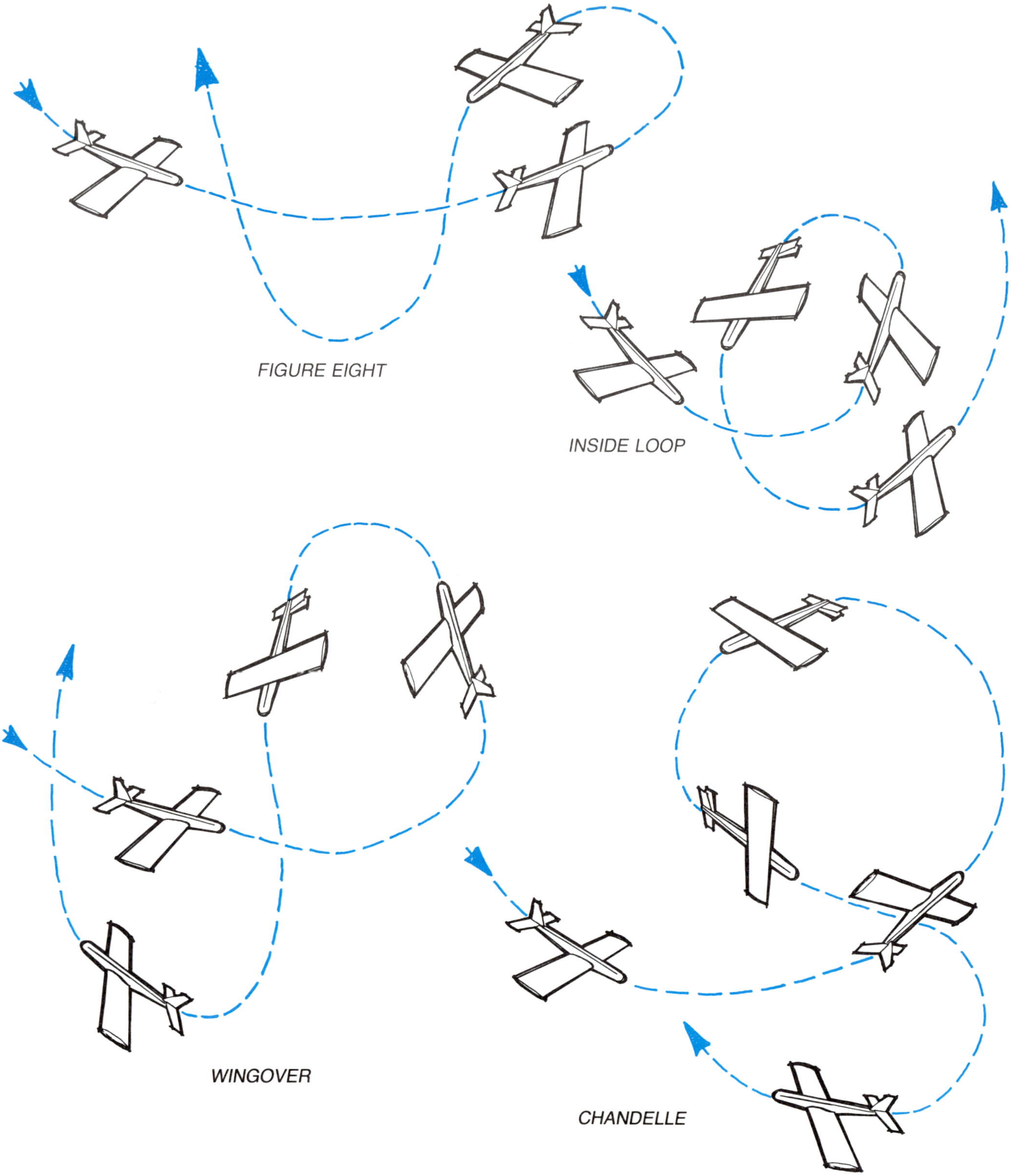

and is upside down, apply a little down elevator and right or left rudder; the racer will turn right side up. Hold the nose down to maintain speed and execute the second half of the eight in exactly the same way as the first half except use opposite rudder.

When you can perform three Cuban Eights in succession, you should have no trouble with any other stunts you can think of, including inverted flight. When flying inverted, elevator controls are reversed but *rudder and aileron controls remain the same as in upright flight.* Practice inverted flight at the highest possible altitude because mistakes at low altitude usually result in a trip to the workbench.

Be aware, too, that slope racers are the fastest models in the sky. An Austrian group made a slope racer, the *Arrow,* go 249 mph. That's almost 50 mph faster than the fastest powered model airplane. That's right; the fastest RC aircraft is a glider!

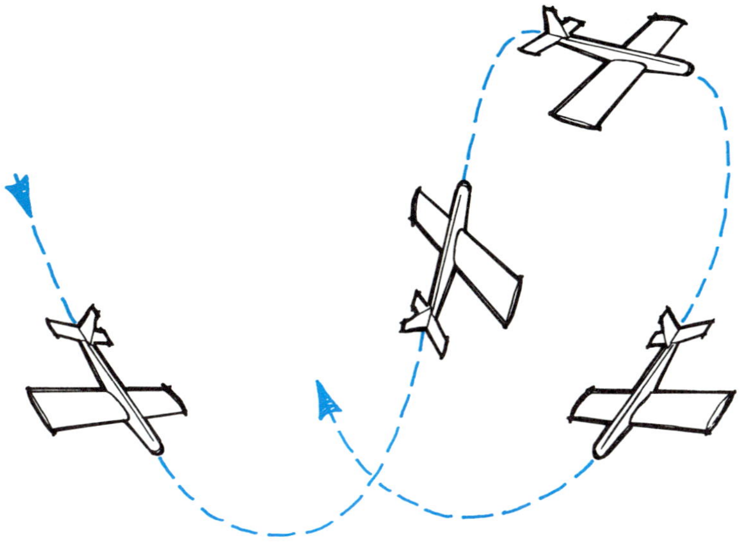

HAMMERHEAD STALL

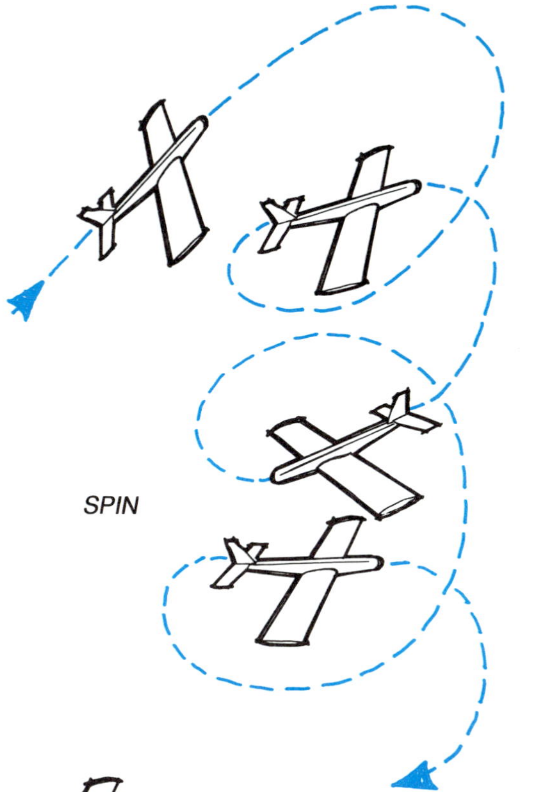

SPIN

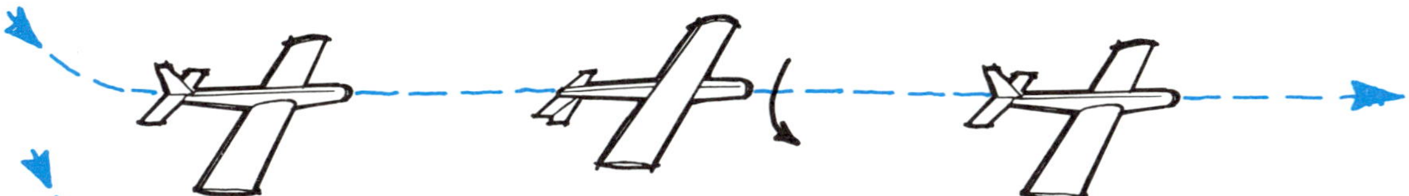

SNAP ROLL

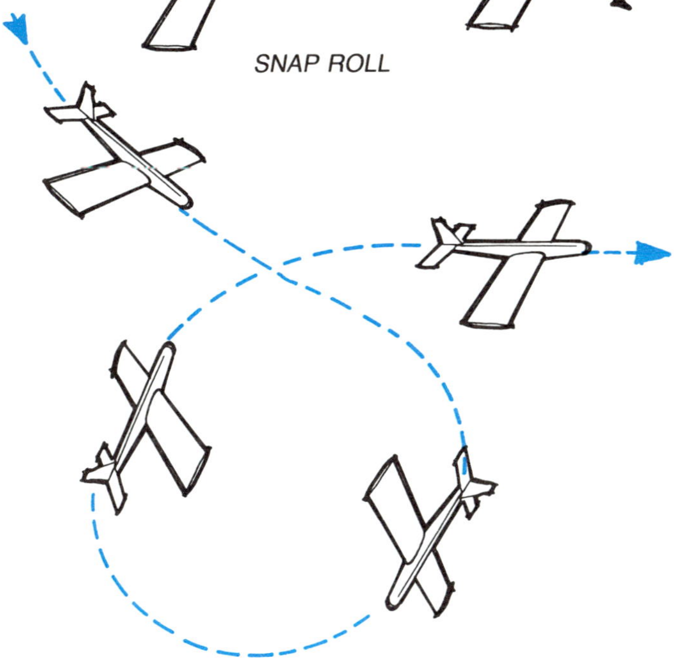

OUTSIDE LOOP

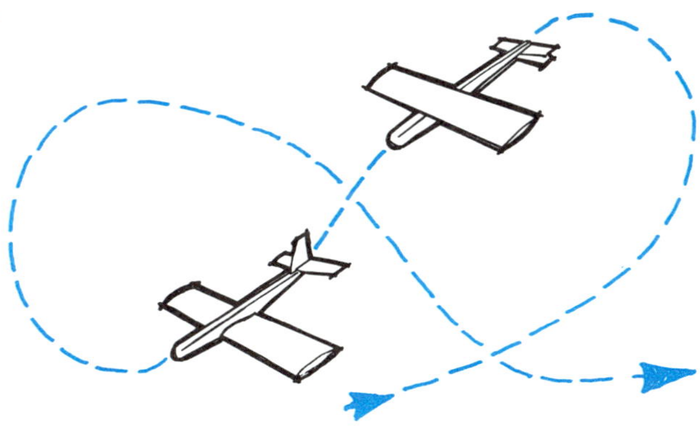

CUBAN EIGHT

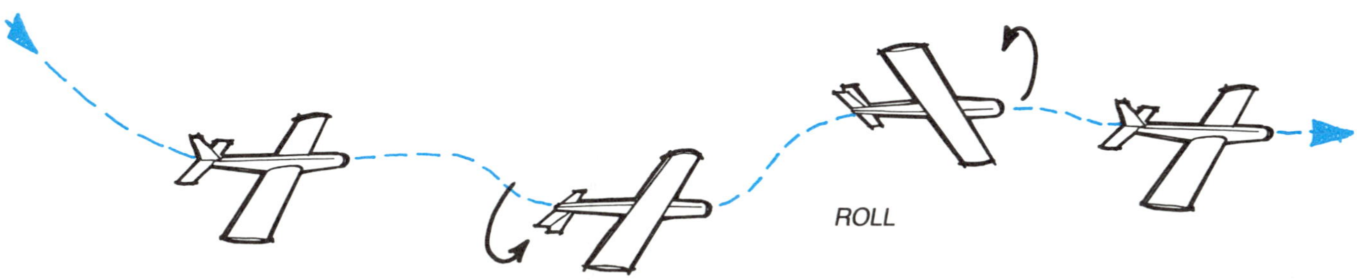

ROLL

Clubs, Contests, and

People sometimes need a little encouragement and as far as flying radio control gliders is concerned, I know that learning alone is a slow process. Mistakes are fewer and less troublesome if there are others to guide you, so I suggest you join an RC club and take part in its activities. Your hobby shop can put you in touch with local clubs; most welcome new members and can provide experienced instructors. Most clubs have access to public or private flying sites.

Almost all RC clubs in the United States are affiliated with the Academy of Model Aeronautics. The AMA, which has more than 72,000 members, sponsors regional and national championships, and represents modelers before the FCC and other government agencies. Annual membership dues are modest considering the $1,000,000 insurance policy and the monthly journal, *Model Aviation*, that is sent to all members. AMA membership is a prerequisite to participation in most contests. You can get a membership application from any of the modeling magazines or directly from the AMA.

The Model Aeronautics Association of Canada (MAAC) offers similar services to Canadian citizens; nearly every other country also has a national model aviation club — get their addresses from your hobby shop. The AMA and the MAAC are in turn members of the Federation Aeronautique Internationale. The FAI is involved worldwide in all flying activities from model aviation to the space program. It keeps the flying world informed on the latest developments in aerodynamic research and certifies record-making flights.

At least one international group, the League of Silent Flight, is solely for RC sailplane pilots. Its goals are to promote RC gliding throughout the world and to recognize individual skills in flying radio control sailplanes. Pilots earn proficiency badges from level one to level five as they perform increasingly difficult flights. If you spot a man or woman with a level five badge you can be sure this pilot has done some serious soaring.

Competition

Competitions among sailplane pilots are based on

(Left) Prizes donated by local hobby shops are displayed before an RC glider contest gets under way. (Right) To prevent interference from models operating on the same frequency, transmitters are impounded until shortly before they are required for flights.

Opportunities

(Left) Between rounds, contestants evaluate glider designs and flying techniques. (Right) A pilot and his helper prepare to launch. Flight officially begins the instant the glider leaves the towline. Motorcycle stands by to retrieve line after launch.

After you have gained proficiency with smaller models, you may want to build a monster. This 1/3-scale racer has more than twice the wingspan of most basic trainer glider models.

two types of criteria. First, there is the standard breakdown by skill level — novice, intermediate, and expert. Then there is a division by glider class:

- Two-meter class. Wingspan less than 78¾″. Any number of controls.
- Standard class. Wingspan 100″ or less. Rudder and elevator controls only.
- Modified standard. Wingspan 100″ or less. Any number of controls.
- Unlimited. Wingspan more than 100″. Any number of controls.

Most contest flights are scored on duration and landing accuracy. Each pilot makes three flights, staying aloft as long as possible during each. He earns extra points by landing within a specified distance of a target. In other events the winner is the pilot whose glider performs the most precise aerobatics or whose sailplane covers a prescribed course most accurately. There are even glider combat events in which the object is to cut a ribbon trailed by the "enemy" glider.

Once you get started in RC gliding you'll find there's no end to the helpful information available. Fellow pilots are handy sources, as are the columns and feature articles in modeling magazines such as *Model Airplane News* and *RC Modeler.* Your public library has or can get several books on full-size and model gliders. Finally, don't ignore the fine British hobby magazines and books.

Challenges

The fun of RC soaring is not over when you have learned to fly your first glider; in fact, the challenges have just begun for you. Space-age materials, exotic glider designs, and cross-country flying with models up to 1/3-scale are just a few of the advanced aspects of our hobby.

Designers have come up with a huge 30-foot wingspan electric-motor assisted glider powered by solar cells built into the upper surface of the wing. This model flies so high it has to be tracked by radar. It's not likely that many hobbyists will build more solar-powered gliders in the near future, but high-performance 1/3-scale models are gaining in popularity. They usually have fiberglass-sheeted wings incorporating Wortmann and Eppler airfoils and boast such features as carbon-fiber main spars, scale Schempp-Hirth spoiler systems, retractable landing gear with closing belly doors, electronic mixing systems which couple ailerons and rudder for perfectly balanced turns, and thermal-sensing telemetry devices. The photo shows my 1/3-scale Glaser-Dirks 15-meter class racer. It's a beautifully designed glider scaled down for RC with five-channel operation: rudder, elevator, ailerons, spoilers, and retractable landing gear. I can couple the ailerons and rudder to one servo and use the extra servo for aero-tow releases.

My glider has yet to be towed up by a powered RC plane — simply because I'm too scared to trust anybody else's skill with my prize possession — but I have witnessed this feat. I was attending a contest for full-size gliders, and after all the pilots had lifted off a couple of European chaps began to assemble their models. One was a 1/3-scale Piper Cub, the other was a 1/3-scale 15-meter class racer. When the Cub's engine was running smoothly they hooked up a 70-foot tow rope to the nose of the sailplane and off they went down the runway. The sailplane lifted off first and retracted its gear. After they were airborne, each plane made three 100-yard circles, climbing to 500 feet. On release the tow plane made a steep diving turn to the left and the sailplane made a graceful climbing turn to the right and flew off looking for thermals. The landing was a showstopper. The glider made a high-speed pass, deployed spoilers to lose speed, came down to 3″ off the deck, retracted the spoilers, and cruised 60 yards at the same incredible 3″ altitude. Just when I was certain the glider would stall, the belly doors popped open, the landing gear extended, the glider touched down, and rolled to a stop 8″ from the pilot's toes.

As you might expect, flying these 1/3-scale ships is more intense than handling smaller models. The stall is more severe and the glider is more likely to spin when it stalls. Because they have 16-foot wingspans and weigh from 10 to 30 pounds, they are awkward on the ground. However, they launch from a hi-start or winch just as easily as a smaller glider. Instead of holding the glider and launching from waist or shoulder height, you set the glider on the ground, hold it by the vertical stabilizer, hook up a hi-start or winch, and let go. The glider rolls down the runway and rises gracefully into the air.

These big birds are made to fly farther faster, so the desert is the best place to fly them. If you don't live near a desert any flat unobstructed area with clear visibility will do. All you need is a road that runs straight for several miles and a pickup truck with padding or a beach chair in the cargo area. A car with a sun roof or a convertible is also fine. To fly cross-country, launch the glider and gain altitude, then get in the car or pickup and follow the glider. When you're ready to search for lift again, stop the car, gain altitude, set the ship back on course, and drive on till the glider starts to come down.

If cross-country flying doesn't appeal to you, perhaps you'll try endurance flying. One pilot has kept his RC glider aloft for almost 28 hours and there's no reason why longer flights can't be made. Or you may want to break the Austrians' speed record. Perhaps you would like to build a superdetailed RC version of your favorite full-size glider. If you're chiefly interested in electronics, there are numerous applications of microcomputers and minicomputers to RC gliding you can investigate. RC gliding can be as relaxing as cruising with a basic two-channel trainer or as exciting as racing a complex 1/3-scale sailplane. Happy flying!

Useful Addresses

Academy of Model Aeronautics

1810 Samuel Morse Drive, Reston, VA 22090. Membership in the AMA includes a $1,000,000 insurance policy, a subscription to *Model Aviation* magazine, and many other benefits.

Radio Control Buyers Guide

Kalmbach Publishing Co., 21027 Crossroads Circle, P.O. Box 1612, Waukesha, WI 53187. This annual contains descriptions of glider kits, RC systems, accessories, and modelbuilding tools.

Department of Communications

473 Albert Street, Ottawa, Ontario K1R 5B4. The DOC issues radio licenses to Canadian RC modelers.

Federal Communications Commission

1919 M St., NW, Washington, D.C. 20544. The FCC issues radio licenses to U.S. RC modelers.

League of Silent Flight

P.O. Box 647, Mundelein, IL 60060. Write the LSF for information about their pilot proficiency rating program.

Model Aeronautics Association of Canada

Box 9, Oakville, Ontario L6J 4Z5. The MAAC is Canada's national model aviation organization.

National Soaring Society

3755 Berkley Lane, Lumberton, North Carolina 28358. In addition to sponsoring RC glider contests, the NSS publishes a monthly magazine, *Sailplane*.

Acknowledgments

Ace R/C, Astro Flight, Cox Hobbies, Craft-Air, Hobby Shack, Kraft Systems, and Pierce Aero provided kits or photographs that helped make this book possible. I thank the members of the Central Valley RC Soaring Club who gave permission to photograph scenes at one of their contests and I thank my students, who pointed out the need for an RC glider pilot's handbook. My wife and daughters helped at each stage of this project — I could not have carried it through without their assistance.

Credits

Photographs by Jack and Mary Schroder, except page 6, Ace R/C, Inc.; pages 7 and 8, Kraft Systems, Inc.; and page 9, Art Schmidt.